T0140953

The Electroluminescent Material Alq_3

Thermal, Structural and Photophysical Properties

Bibliografische Information Der Deutschen Bibliothek

Die Deutsche Bibliothek verzeichnet diese Publikation in der Deutschen Nationalbibliografie; detaillierte bibliografische Daten sind im Internet über http://dnb.ddb.de abrufbar.

ISBN 3-8325-0494-X

Logos Verlag Berlin
Comeniushof, Gubener Str. 47,
10243 Berlin
Tel.: +49 030 42 85 10 90
Fax: +49 030 42 85 10 92
INTERNET: http://www.logos-verlag.de

Für Robin

Contents

1 Introduction

The material tris(8-hydroxyquinoline)aluminum(III) (Alq_3) shown in Figure 1.1 has been used for more than 40 years for research purposes [Ohn59] and it is the most frequently used material for organic devices based on small molecules. Since 1987 it has been the standard material for organic light emitting devices (OLEDs) and since 1997 it has been the main material used in commercially available organic displays[1]. However, so far comparatively few investigations have been devoted to the fundamental properties of this material. This surprising fact was the stimulus for the research presented here. Knowing more about Alq_3 should help to obtain more basic insight into the working mechanism of these devices and optimize the preparation conditions which may create more efficient displays.

Increasing interest in this material for technical applications started with the report on efficient electroluminescent devices using Alq_3 as the active medium [Tan87]. These so-called OLEDs opened the way for a new display generation. After nearly two decades of intensive research and development of OLEDs Alq_3

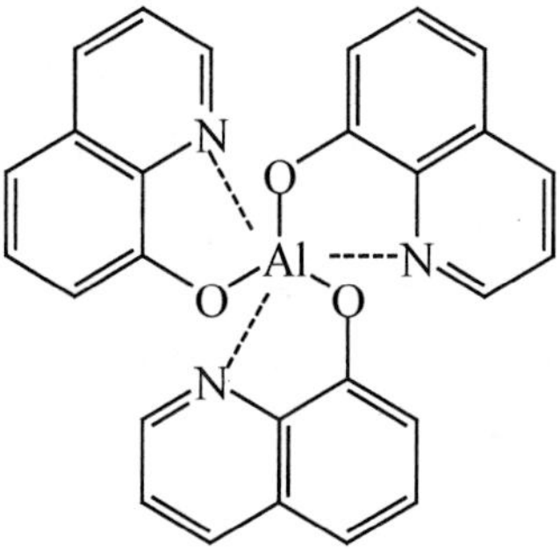

Figure 1.1: *Chemical structure of Tris(8-hydroxyquinoline)aluminum(III) (Alq_3)*

[1] In November 1997 Pioneer Co. in Japan commercialized a monochrome 256x64 dot matrix OLED display for automotive applications.

still continues to be the workhorse in low-molecular weight materials for these devices. It is used as electron transporting layer, as emission layer, where green light emission is generated by electron-hole recombination in Alq_3, and it also serves as host material for various dyes to tune the emission color from green to red [Tan89]. Many studies in this field have focused on the optimization of device characteristics with respect to efficiency and long-term stability or on the understanding of charge transport properties of amorphous thin films [Shi97, Hun97, Azi99, Kub00, Tsu99, Bur96, Bar99, Stö00, Brü01]. However, until recently comparatively few investigations have been devoted to the material's electronic and optical properties - in particular in the crystalline state - as well as to the dependence of these properties on the preparation conditions [Bri00, Bra01, Cöl02, Cöl03]. This is very surprising; since the first publication on OLEDs based on thin films it has been known that the so-called "amorphous" film of Alq_3 has nanocrystalline domains [Tan87], which raises many questions concerning the morphology and properties of Alq_3. For example, what kind of crystalline phases can be formed by Alq_3 and what are their electrooptical properties? What is the packing of the molecules? Packing and intermolecular interactions are important for optical properties as well as for their electrical characteristics and the transport mechanism of charge carriers.

Another unresolved issue concerns the isomerism of the Alq_3 molecule. It is well-known that octahedral complexes of the type MN_3O_3, where M is a trivalent metal and N and O stand for the nitrogen and oxygen atoms in the quinoline ligands, can occur in two different geometric isomers: meridional and facial, as shown in Figure 2.1 [Kau74]. Nevertheless, only the meridional isomer has been clearly identified and no direct experimental evidence for the facial isomer has been found so far. Therefore it was generally believed that the meridional isomer is predominant, both in amorphous films and crystals of Alq_3. The existence and the properties of the facial isomer are discussed in detail in the literature and a key issue is its possible presence in sublimed Alq_3 films [Lar68, Bak68, Maj70, Hal98, Joh99, Mar00, Kus00, Bri00, Ich02, Cur02]. Many suggestions have been made about its influence on trap density, charge carrier transport and thus on the characteristics and performance of OLEDs. For example the higher dipole moment of the facial isomer is expected to influence the morphology of the film as well as the injection of charge carriers at the interface. In addition the different HOMO and LUMO-levels predicted for the facial isomer are seen as influencing the injection barrier and acting as traps for charge carriers [Cur98, Hum00, Mar00, Ste02, Ito02, Ama02a]. Therefore the question is whether the facial isomer is present in one or the other modification of Alq_3 or not and, in the former case, if it is possible to isolate it. The isolation of the facial isomer is of great interest, as it will allow us to examine its properties separately and thus to clarify its role in OLEDs.

The purpose of the research summarized in this book was to address these open questions by investigation of (poly)crystalline Alq_3 and thus to obtain more insight into its properties. To this end a train sublimation method with a temperature gradient inside a glass tube was used, giving a separation of fractions with different characteristics, and this made it possible to identify new crystalline phases of Alq_3. Independently, a group in Italy has begun to investigate crystalline Alq_3, also identifying different polycrystalline phases (the so-called α-, β- and γ-phases) including structural analysis. Their publication is very helpful for this work and is discussed in the relevant chapters [Bri00].

The temperature gradient in the sublimation tube greatly affects the formation of the different phases and thus studies on the thermal properties of Alq_3 have been carried out. From these it was determined how these phases form, how to transform the phases into each other and how to obtain samples of pure phases, which were the basis for further investigations.

Photoluminescence and photoexcitation measurements of the Alq_3-phases demonstrate significantly different properties of the phases, the vibronic structure was resolved and the intersystem crossing rate determined. Furthermore the very first experimental data of the triplet state in Alq_3 were measured and compared for the different phases. It was also shown that the phases can be clearly distinguished by using vibrational analysis, as IR spectroscopy gives clear fingerprints for each phase.

Furthermore the measurements were used to address the important question of isomerism, in particular the existence of the facial isomer. In fact, all experimental results on the new blue luminescent δ-phase are in good agreement with theoretical calculations made for the facial isomer of Alq_3. Moreover, high resolution X-ray powder diffraction measurements with structural analysis including Rietveld refinement confirmed that the facial isomer constitutes the δ-phase of Alq_3.

This book specifies the thermal, structural and photophysical characteristics of different polycrystalline Alq_3-phases. In particular the new blue luminescent δ-phase is shown to contain the facial isomer. For the first time the existence of this isomer is clearly demonstrated, and this is expected to play an important role in thin Alq_3-films used in OLEDs. From the results of this work it is now possible to obtain the pure facial isomer of Alq_3 in large quantities, providing the basis for further investigations to determine its effects on the performance of OLEDs.

2 The Material Alq_3 and its Applications

This Chapter gives a brief overview of the known molecular and solid state properties of the material and its applications. In the first section the general molecular properties, especially the isomerism of the molecule, are discussed. The second part focuses on evaporated films of Alq_3 mainly used for electrical applications like organic light emitting devices (OLEDs). The basic working mechanisms of OLEDs are introduced and some basic electrical properties are summarized.

2.1 General and Molecular Properties

Tris(8-hydroxyquinoline)aluminum(III) ($Al(C_9H_6ON)_3$), commonly named Alq_3, is an organometallic complex with a central Al atom and three ligands of 8-hydoxyquinoline (8-Hq) shown in Figure 1.1. The synthesis of the material used for this research is described in the appendix. For purification the substance is in general dissolved and re-precipitated several times in chloroform, followed by train sublimation (see experimental methods in Chapter 3).

In Alq_3, the Al^{3+} ion stabilizes the molecular orbitals of the 8-hydroxyquinoline anion by means of interaction between the metal ion and the oxygen that carries most of the electron density of the ligand [Sug98]. The nitrogen-aluminum interaction is relatively weak and a so-called coordination bond. The material is stable under ambient conditions in the dark. However, instability and degradation processes are observed due to photooxidation as well as in electrically driven devices due to reactions with oxygen and water [Ngu98, Pap98, Azi98, Azi99, Sch01].

Alq_3 belongs to the class of metal chelate complexes. The process of ring formation as a result of ligands around a central metal atom is known as chelation and was discovered in 1893 by the German scientist, Werner. The word "chelate", which describes the ring, was originally proposed in 1920 by Margan and Drew. It is derived from the Greek "chele", meaning a lobster's claw.

Chelate complexes are used for a large variety of different applications in many scientific fields ranging from biology and medicine to technical applications [Bro73, Wil72]. Very recently there has been new interest in the synthesis of new chelate complexes including a heavy central metal atom (e.g. Ir or Pt) for display applications because they allow emission from both singlet and triplet states, which leads to very efficient OLEDs with internal quantum efficiencies close to 100% [Ada01, Wat01].

This variety of different applications is also true for Alq_3, for example in medicine for treating tumors, infections and Alzheimer's disease, in biology for stimulation of plant growth, and in industry for corrosion protection [Gee85, Mur89, Leh91, Mon91, Mar01]. It also was used for many years in analytical chemistry for a gravimetric determination of various metal cations in solution [Ohn59, Cha67]. Now it is the most frequently used low-molecular weight material for OLEDs.

Werner's classical research on coordination compounds established the occurrence of several different types of isomerism – the existence of two or more compounds of the same empirical formula but with different arrangements of atoms within the molecule [Kau74]. In general one distinguishes between optical isomerism and geometric isomerism. In Alq_3 two geometric isomers exist, namely the facial and the meridional isomer. Facial isomers and meridional isomers of Alq_3 can each exist in two different forms, their optical isomers. The different isomers are shown in Figure 2.1. Possible conversion processes of the two geometric isomers of Alq_3 and its optical isomers are discussed in a theoretical paper by Amati et al. [Ama02].

Optical isomerism is found in coordination chemistry whenever a molecular structure is such that two forms, related as an "object" and its "mirror-image", exist. The isomers are distinguishable by their effect on linearly polarized light. In general one isomer, the "object", rotates the plane of polarization in one direction, whilst the other, the "mirror-image", rotates it in the opposite direction. The two optically active isomers are called enantiomers and they are described as being enantiomorphous with each other. The two enantiomorphous configurations Δ and Λ of Alq_3 are shown in Figure 2.1. The enantiomers are related as object and mirror-image, and neither can be superimposed on the other.

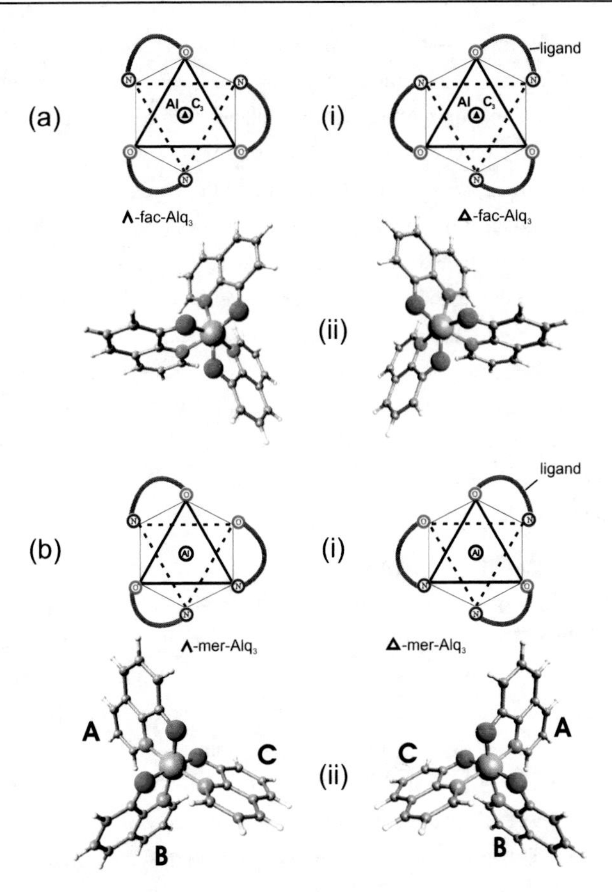

Figure 2.1: *Structure of the facial (a) and the meridional (b) isomer of Alq3 and their enantiomers shown in a schematic representation (i) and a 3D model (ii). In (i) the position of the organic ligands is only shown schematically by gray bars. For the meridional isomer the labeling of the ligands by A, B and C is given*

Geometric isomerism represents different spatial distributions of a given set of atoms or groups around a central atom. In octahedral chelates such as Alq_3, geometric isomerism exists because of different arrangements of the unsymmetrical ligand molecules in coordination with the metal. These, illustrated in Figure 2.1, are designated facial (fac) and meridional (mer). In the facial isomer, the three O-atoms are situated at the corners of one triangular face of the octahedron and the three N-atoms are at the corners of another face parallel to the first. In the meridional isomer, the O atoms lie in a plane which is perpendicular to a similar plane containing N-atoms. Both planes include the central Al-atom. The two geometric isomers are clearly distinguished by their symmetry. In the facial isomer, all three ligands are equivalent and the complex has a C_3 rotation axis. In the meridional form, the ligands are in different positions relative to the atoms of the adjacent ligand and the molecule has a reduced symmetry. The distinction between the ligands is clearly seen in relation to the O and N atoms of the other two ligands in the trans-positions. (In the facial isomer all O atoms are opposite the N atoms, which is not the case in the meridional form). This makes it possible to label each ligand of meridional Alq_3 by A, B and C respectively, as shown in Figure 2.1(b). These differences in symmetry influence the vibrational modes of the molecule and can be analyzed by infrared (IR) spectroscopy, as will be discussed in Chapter 4.4.

From this geometry it is clear that the two stereoisomers are likely to have significantly different dipole moments. With the MOPAC software we calculated 4.3D (D: Debye) for the meridional and 8.1D for the facial isomer. In the meridional isomer, the dipole is oriented toward the oxygen atom, which has a nitrogen atom opposite. In the facial isomer, it points to the center of the triangular face defined by the three oxygen atoms. Similar results of other theoretical calculations have been published (e.g. in Ref [Cur98] meridional: 4.3D facial: 7.9D) and in these the meridional isomer was found to be more stable than the facial isomer by approximately 17 kJ/mol [Cur98, Mar00].

Due to their different symmetry the geometric isomers are expected to have different HOMO and LUMO levels. Figure 2.2 shows theoretical results of Amati et al. using time dependent density functional theory (TD-DFT). In Figure 2.2(a) molecular orbital energies of both isomers are shown. The molecular orbitals can be grouped in sets of three closely-spaced orbitals. A number from I to IV labels each set. All the molecular orbitals are mainly localized on the chelants. Furthermore, each molecular orbital of a set can be considered to be a linear combination involving the fragment orbitals shown in Figure 2.2(b) and labeled by the same letter of the corresponding set. Set-I molecular orbitals are substantially a linear combination of three fragment orbitals of type I with each fragment orbital localized on each chelant. Although this is common to both the isomers, there are differences in the delocalization. In meridional Alq_3, each molecular orbital contained in each set can be attributed to a single chelant and thus – as mentioned above – be named using the letters A, B and C. In facial Alq_3, due to the C_3 point symmetry, the molecular orbitals are equally distributed on the three chelants. For this reason the sets are labeled by the symbol of the irreducible representation to which they belong ("e" for the degenerated doublet and "a", respectively). For both isomers, set-I and set-II orbitals are occupied, set-III and set-IV orbitals are unoccupied. The HOMOs (set-II) are localized mainly on the phenoxide side of the ligands and the LUMOs mainly on the pyridyl side. LUMO + 1 orbitals (set-IV) are more uniformly distributed on the ligand. Similar results have also been reported by Curioni et al. using ab initio calculations [Cur98]. As a result of these calculations, the distance between HOMO and LUMO level is different for the geometric isomers. The LUMO-HOMO gap is predicted to be larger in the facial isomer by about 0.13eV to 0.3eV [Ama02a, Cur98]. It is expected that this will result in different optical properties that can be probed by measuring photoluminescence and absorption spectra (see Chapter 5).

Further theoretical work has been carried out to understand the effects of a

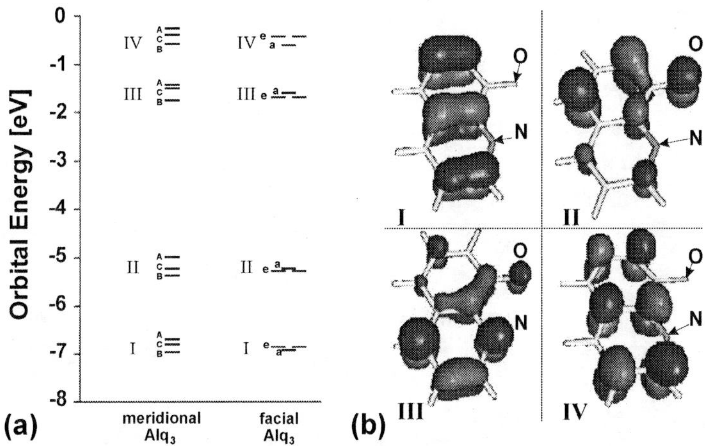

Figure 2.2: *(a) Computed energies of molecular orbitals for meridional Alq_3 and facial Alq_3. Set-I and set-II orbitals are occupied, set-III and set-IV orbitals are unoccupied. In meridional Alq_3 the molecular orbitals can be attributed to a single chelant and are labeled by A, B, and C respectively. In facial Alq_3 the orbitals are labeled by the symbol of the irreducible representation to which they belong. Figure (b) shows fragment orbitals involved in the molecular orbitals calculated in (a). The HOMOs (set-II) are localized mainly on the phenoxide side of the ligands and the LUMOs (set-III) mainly on the pyridyl side [Ama02a].*

positive or negative charge on the molecule, considering both molecular ions and their atomic configurations [Cur98, Bur96, Joh99, Yan02]. The ionization potential, the energy to create a radical cation, is in the range of 5.6-6.0eV, as also determined by UPS measurements [Cur98, Ham93, Sch95, Pro97]. For both isomers relatively small energy changes associated with structural relaxation are reported, and the molecule exhibits a relatively high structural stability against injection of one charge, whether positive or negative. However, recent experiments on devices show that the Alq_3-cations involved in hole transport are unstable and may contribute to photoluminescence and electroluminescence aging in OLEDs [Pap96, Mat97, And98, Azi99, Cho99, Pop01]. In the meridional isomer, owing to the lack of symmetry, the extra charges are not localized uniformly on the three ligands: the hole has a larger amplitude on ligand A and the electron on ligand B. The energy gain of the excess electron due to structural relaxation of the molecule was estimated to be 0.06eV and 0.13eV in the meridional and the facial molecule, respectively [Cur98].

Until now the facial isomer of Alq_3 has only been predicted theoretically. Its different geometry, higher dipole moment and different electronic properties are expected to be of great relevance for film forming and transport properties and thus for the performance of OLEDs in display applications. For many years several groups have been trying to find the facial isomer and much research has been done on this topic. However, only the meridional isomer has been clearly identified and thus it was generally believed that Alq_3 only consists of the meridional isomer [Lar68, Bak68, Maj70, Sch91, Fuj96, Bur96, Hal98, Mar00, Kus00, Bri00, Ich01, Cur02].

In this work the first clear evidence for the existence of the facial Alq_3 molecule is given. The facial isomer is isolated and characterized for the first time.

2.2 Thin Films of Alq_3 and Their Properties in Devices

The material Alq_3, whose general and molecular properties are described in the previous section, is mostly used as thin films in OLEDs. Although this book focus on Alq_3 in its crystalline state, a summary of the principal characteristics and properties of these thin Alq_3-films is given in this section, as they are of great relevance for applications. Alq_3 forms homogenous transparent thin films upon vacuum deposition and it has a relatively high glass transition temperature T_g of 175°C [Nai93]. Although these films are commonly called "amorphous", they are known to contain nanocrystalline regions causing small structural defects [Tan87, Jon00]. This explains why it is crucial to learn more about the structural properties of Alq_3 in OLEDs.

The presence of the different isomers of Alq_3 is believed to have a strong effect on film-forming properties and long-term stability in the glassy form as they should act as a stabilizing factor in the amorphous phase by preventing recrystallization, which appears to be a prerequisite for good performance and thermal stability of the device [Kim95, Bur96, Ham97, Che98, Cur98, Bri00]: Building up an extended crystalline arrangement of Alq_3 molecules would require a precise ordering of the two corresponding enantiomers and for this sufficient molecular mobility is necessary. However, molecular mobility is strongly hindered by the existence of high energetic barriers due to the dipole-dipole interactions between Alq_3 molecules (as mentioned above, both geometric isomers have a strong dipole moment: meridional: 4.3D, facial: 7.9D) The intrinsic polymorphism of Alq_3 (see Chapter 4.1) as well as local fluctuations in the concentration of the geometric isomers and its enantiomers during evaporation, combined with strong dipolar interactions between Alq_3 molecules, could constitute favorable conditions for obtaining stable glassy amorphous films after sublimation.

In order to measure the electrical properties of Alq_3, such as the mobilities of electrons and holes, the mechanism of charge transport as well as the distribution, depth and influence of traps, it is necessary to make electrical contacts to the material. As Alq_3 crystals obtained so far have been very small (in general less than 500μm, see Chapter 4.1), no electrical experiments on single crystals have been possible [Kaw01, Bri00]. On the other hand it is quite easy to make contacts to films, simply by thermal evaporation of metal electrodes.

As shown schematically in Figure 2.3, electroluminescence in organic solids requires several steps, namely injection ①, transport ②, capture ③ and radiative recombination ④ of positive and negative charge carriers inside an organic layer with a suitable energy gap to yield visible light output. To describe these OLEDs it is common to use concepts derived from inorganic semiconductor physics, but one should be aware of the differences of organic semiconductors to their inorganic counterparts. It is also not possible to simply adopt mechanisms developed for molecular crystals, because the films mainly have a disordered amorphous character. In the simple scheme in Figure 2.3 the spatial variation of the molecular energy levels is drawn in a band-like manner. However, one has to bear in mind that these organic semiconductors are disordered materials. The energy levels of charged molecules are different from the neutral state levels due to structural relaxation, but for reasons of simplicity these polaronic effects are not included in this figure. Furthermore, compared to inorganic semiconductors, most of the materials in OLEDs are wide-gap materials with energy gaps of 2-3eV or even more (band gap of Alq_3: 2.7eV, HOMO: 5.7eV; LUMO: 3.0eV) [Bur96, Pro97]. Therefore the intrinsic concentration of thermally generated free carriers is negligible ($<10^{10}cm^{-3}$) and from this point of view organic materials used in OLEDs can be considered more as insulators than as semiconductors. Unlike in inorganic semiconductors, impurities usually act as traps rather than as sources of extrinsic mobile charge carriers. Nevertheless, this simplified scheme in Figure 2.3 is helpful to understand the principal mechanisms of these devices.

Once carriers are injected into the organic material these are transported in the electric field towards the counter-electrode (Figure 2.3b ① ②). Due to disorder charge carrier transport in organic materials can be considered as hopping between sites with different energy and distance. Additionally, carriers can be intermittently trapped in gap states resulting for instance from impurities or structural defects. This results in low carrier mobilities, which are typically between 10^{-3} and 10^{-7} cm^2/Vs at room temperature and in many cases significantly depend on temperature and the magnitude of the applied electric field [Bor93]. The field dependence (F: field) can be described according to

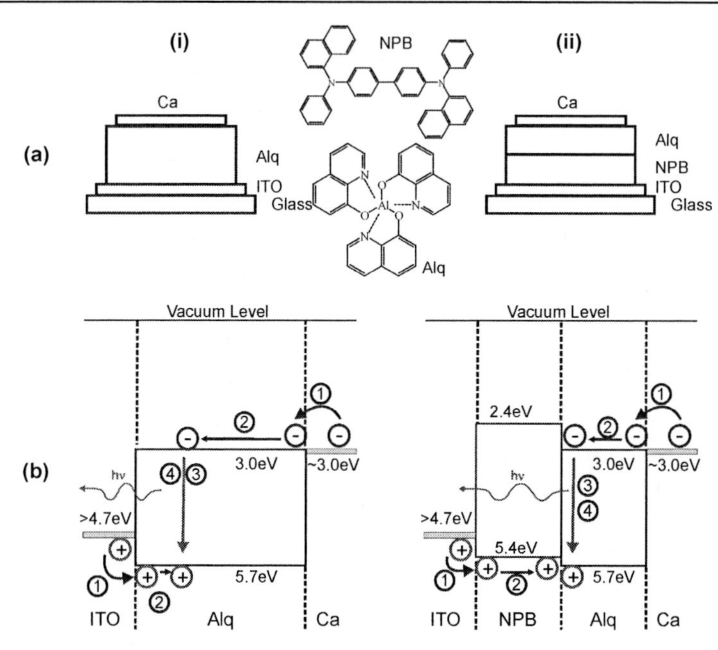

Figure 2.3: *a): Typical structures of single (i) and double (ii) layer devices. The organic layers of Alq_3 and NPB respectively are sandwiched between two electrodes of indium tin oxide (ITO) and calcium (Ca). The structures of the organic materials are also shown. b): Schematic energy level diagram and principal working mechanism of a single (i) and double layer (ii) device. ①: injection of the charge, ②: charge transport, ③ formation of a neutral excited state (exciton) ④: radiative recombination and light emission.*

$\mu(F) = \mu_0 \exp(\beta\sqrt{F})$ with decreasing zero-field mobility μ_0 and increasing β-factor for lower temperature. Measured values for electron mobility in Alq_3 are in the range between 10^{-5} to 10^{-8} cm^2/Vs [Ber01, Kep95]. Hole drift mobility in Alq_3 at 300K is almost three orders of magnitude less at low fields compared to electron drift mobility. However, with increasing electric field the difference in the mobilities of both charge carrier types becomes smaller due to a stronger dependence of hole-drift mobility on the electric field [Kep95, Tsu98, Nak99].

For the mechanisms of charge transport ② several models such as trapped-charge-limited current, hopping transport and space-charge-limited current have been proposed [Bur96, Ber01]. Furthermore, Alq_3 has the unusual property that transport for the carrier species with the much higher mobility (electrons) is dispersive while it is non-dispersive for the lower mobility carriers (holes) [Nak99]. More detailed investigations on mobility and transport mechanism in Alq_3 can be found in Refs [Ber01] and [Brü01].

The depth, concentration and distribution of traps also play a role in the models for charge transport. Their depth has been determined with methods like thermally stimulated luminescence [For98], thermally stimulated current [Ste02] and from electrical characteristics [Brü01, Bur96, Jeo02, Ber01] to be between 0.10 – 0.25eV, with a mean trap depth of about 0.15eV. The density of traps has been estimated to be of the order of $10^{17}cm^{-3}$ [Ber01]. The origin of the trap states has been attributed to impurities, structural defects and the presence of the two geometric isomers. However, no responsible impurity has been identified so far and thus the latter two points seem to be dominant.

For electroluminescence charge carriers of opposite sign have to recombine and form an exciton③, which then decays radiatively ④ (Figure 2.3b). Due to low carrier mobility the process of electron-hole capture is diffusion-controlled and is therefore of the Langevin type [Alb95a, Alb95b]. Moreover, the singlet exciton may not only decay radiatively but also nonradiatively. The ratio of radiative decay of singlet excitons in a singlet emitter like Alq_3 is given by the photoluminescence quantum yield. Furthermore, the excitations generated by carrier recombination of nongeminate pairs can be a triplet or a singlet state with a branching ratio of 3:1, setting an upper limit of 25% for internal quantum efficiency, meaning the conversion of injected carriers into photons via singlet excitons. The photoluminescence quantum efficiency (PL-QE) of Alq_3-films was measured in this work to be up to 19%, similar to values reported in the literature [Gar96, Rav03]. Consequently internal QE in Alq_3-devices is limited to less than 5%. Outcoupling of the generated light further reduces the efficiency; thus the external QE of Alq_3-devices is in the order of 2%. As mentioned above, new metal chelate complexes with a high yield of short-lived phosphorescence triplet emission have recently been used to increase the quantum efficiency and to produce very efficient OLEDs [Bal99, Tsu99, Ada01, Wat01].

Chelate complexes such as Alq_3 have been investigated for many years and a multitude of groups are working on Alq_3. It is the most frequently used material in OLEDs. However, there are still many open questions about its properties, knowledge of which might be helpful to learn more about the working mechanisms in OLEDs, and it is very surprising that when this research was first begun only very little was known about its crystalline phases. The different phases of Alq_3, their crystalline as well as their molecular structure, and their specific properties are discussed in Chapter 4 and Chapter 5.

3 Experimental Methods

This chapter describes the experimental methods that were used for preparation and characterization of samples. Furthermore, a cryostat system was setup during this work for performing electrical and optical measurements both on films as well as on polycrystalline powders in the temperature range between 6K and 300K. This setup consists of a continuous-flow liquid helium cryostat (Oxford Instruments CF200) with four optical windows. It is equipped with

- different light sources (a He-Cd-Laser , UV-LEDs with interference filters (375nm, 440nm), a halogen light source with a highly stable current source (Heinzinger LNG 50-10), and a tungsten-halogen lamp CVI AS220 (300-2500nm) coupled directly to the monochromator)
- light detection systems (CCD (Princeton Instruments), monochromator (CVI Digikröm DK240), photomultiplier, photodiode, amplifier (Gigahertz Optic P-9202 and EG&G 5182), lock-in (Stanford Research SR850DSP))
- electrical instrumentation (Keithley SMU 236, HP 3245A).

Its versatility allows a variety of different experiments to be performed as will be described below:

With this setup temperature-dependent photoluminescence (PL) and electroluminescence (EL) spectra can be recorded either with a CCD camera coupled to a spectrograph or using a monochromator in combination with a photomultiplier or photodiode. The CCD is very convenient for measurements of spectra with low intensity or integration of spectra over a certain period of time, whereas the monochromator allows a high resolution of the measured spectrum or the selection of a specific wavelength. With the combination of a light source, chopper, monochromator, photomultiplier (PM), and oscilloscope measurements of the transient PL can be performed. Further, it is possible to measure photoinduced absorption. This is a typical pump-probe experiment, where the sample is pumped by a chopped laser beam and the signal is detected via the monochromator, photodiode and lock-in amplifier. This method was used for investigations on subgap states in Alq_3 that will be published elsewhere.

Moreover, by using two synchronized choppers, this setup is ideal for measurements of the delayed luminescence. The spectra can be recorded with the CCD and for the transient characteristics the monochromator, PM and oscilloscope are used. Furthermore, the setup for measuring the PL-quantum efficiency (PL-QE) is also included.

The sample holder was constructed to allow electric contacts to the samples. Thus, with an appropriate light source (CVI AS220), this setup was used for photocurrent measurements. In addition, measurements to investigate the temperature-dependent characteristics of OLEDs (I-V-characteristics and temperature-dependent EL quantum efficiency) were performed with this setup. Most of the measurements are controlled by a computer and the programs were made with Labview.

This setup was used in particular to address the properties of the polycrystalline Alq_3-phases, namely by measurements of room temperature and low temperature PL-spectra, transient PL, delayed luminescence (spectra and transient decay), and PL-QE.

Characterizing materials like Alq_3 is an interdisciplinary goal combining methods from materials science, chemistry and physics. This led to intensive cooperation for structural analysis of the samples via X-ray diffraction with the Department of Inorganic Chemistry I at the University of Bayreuth as well as with Dr. habil. R. Dinnebier from the MPI in Stuttgart. Another very fruitful cooperation was with the Institute of Physics III at the University of Stuttgart, which had the ideal setup for taking the PL-excitation measurements and where it was possible to measure optically detected magnetic resonance (ODMR) at zero field.

As far as necessary for a full understanding of the data and results presented, the principle of the methods and the information obtained thereby are directly discussed in the relevant chapters. Only the results of the ODMR measurements are given in Chapter 5 without further explanation, and therefore a brief introduction is included at the end of this chapter.

3.1 Sample Preparation

The samples were prepared either by using train sublimation or thermal evaporation if not marked otherwise.

Train Sublimation

The polycrystalline samples of Alq_3 were obtained by train sublimation in a horizontal glass tube (length 90cm, width 1cm) as follows. First about 1.5g of the Alq_3 source material, synthesized as described in the appendix, was filled in a small glass tube (width 8mm) under ambient conditions. This tube was placed at the end of the larger glass tube with its open side to the closed side of the larger tube (see Figure 3.1). Heating of the samples was carried out in a high temperature oven (GERO, SR70-500). During the whole following procedure the tube was evacuated to a pressure of 10^{-2}mbar by a rotary pump. After half an hour in this vacuum the temperature was increased in three steps over about 4 hours until 400°C was reached. The sublimation was then continued for another 4 hours at this temperature. Then, the oven was turned off. At temperatures below 200°C the tube was taken out of the oven and was allowed to cool down to room temperature (for several hours). After this sublimation process, three zones with different crystalline material in the glass tube were observed, as shown in Figure 4.1. A small amount of brown flaky material was left as a residue in the small glass tube in which the source material was filled at the start of this

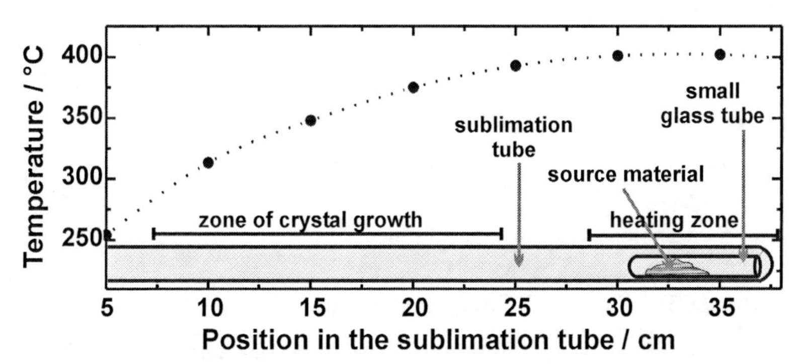

Figure 3.1: *Temperature gradient in the sublimation tube. The heating zone with the small glass tube, which contains the source material, and the zone of crystal growth are also shown.*

sublimation procedure. The temperature gradient along the sublimation tube is shown in Figure 3.1.

Thermal Evaporation

The focus of this work is on the polycrystalline phases of Alq_3. In addition, in order to assess the influence of the morphology, evaporated films of Alq_3 are used. These amorphous films with a thickness between 10nm and 5μm were fabricated on purified quartz substrates at room temperature by thermal evaporation in a high vacuum system (Leybold) at a pressure of about 10^{-6}mbar and a deposition rate of about 10Ås^{-1}.

3.2 Characterization Methods

Differential Scanning Calorimetry DSC

Thermal analysis was performed by differential scanning calorimetry (DSC) using a Netzsch DSC 200. Samples (10-15mg) of polycrystalline powder were placed in aluminum pans under inert atmosphere and heated at a rate of 2°Cmin^{-1} unless otherwise stated. Indium and zinc metals were used as the temperature standard.

X-Ray Diffraction

For crystal structure determination of the different Alq_3-phases two methods have been used. The unit cell of the different Alq_3-phases as well as the influence of preparation conditions on the crystalline structure of the samples were measured at the Department of Inorganic Chemistry I of Prof. Hillebrecht in Bayreuth using $CuK_{\alpha 1}$ radiation, whereas the molecular structure was determined using the synchrotron light source in Brookhaven.

X-Ray Diffraction Using $CuK_{\alpha 1}$ Radiation

X-ray powder diffractograms were obtained with Ge-monochromated $CuK_{\alpha 1}$ radiation (λ=1.54056Å) on a transmission powder diffractometer in Debye-Scherrer geometry (Siemens D50000) as well as on a STOE STADI P transmission powder diffractometer. Data were collected at room temperature in the 2Θ-range from 5° to 35°.

High Resolution X-Ray Powder Diffraction and Refinement of the Molecular Structure

High resolution X-ray measurements as well as specific software for simulation of the structure are necessary to solve the structure of Alq_3 since it is only available as powder. Dr. habil. Robert Dinnebier from the MPI for Solid State Research in Stuttgart, an expert in X-ray powder diffraction measurements for refinements of molecular structures, made the measurements and determined the structure. Details of the experimental procedure and the software used can be found in Ref. [Cöl02].

Fourier Transform Infrared Spectroscopy (FT-IR)

The samples were measured on KBr pellets in a Bruker IFS55 Fourier transform infrared (FT-IR) spectrometer with KBr beamsplitters and a resolution of $2cm^{-1}$. Below $420cm^{-1}$ the KBr optic reduces the quality of the spectrum; thus, to prove the interpretation below $420cm^{-1}$, a Bruker IFS 66v spectrometer with mylar coated beamsplitters, CsI pellets and a resolution of $1cm^{-1}$ was used for the low energy region between $370cm^{-1}$ and $480cm^{-1}$. All spectra shown are raw data, without any further correction (smoothing, background, etc.). In the case of the polycrystalline samples scattering due to crystallinity leads to a broad background with sometimes asymmetric peaks. After comparison with measurements on fine powder it was verified that this does not significantly influence the peak positions.

Photoluminescence Quantum Efficiency (PL-QE)

The photoluminescence quantum efficiency (PL-QE) was measured using an integrating sphere (Bentham IS4, Ø:10cm) and a spectrograph (SOLAR S-380, gratings: 150 lines/mm, 200 lines/mm, and 300 lines/mm) coupled with a liquid nitrogen-cooled CCD detector (Princeton Instruments). The excitation was at 375nm from an UV-LED (Nichia) with an interference filter. An integrating sphere is a hollow sphere which has its inner surface coated with a diffusely reflecting material (barium sulfate or e.g. color from Eastman Kodak No. 6080, reflectivity of >98% over the range from 300nm to 1200nm). When a light source is placed in an ideal integrating sphere, the light is distributed isotropically over the interior surface of the sphere regardless of the angular dependence of the emission. This assumption of an ideal integrating sphere is the limiting factor for the accuracy of the values obtained for the PL-QE. The measurement procedure and the data analysis were performed using the optical scheme and the notations of de Mello et al. [Mel97], which have been optimized by us as described by Tzolov et al. [Tzo01a].

Photoluminescence Emission (PL) and Photoluminescence Excitation (PLE) Measurements

For photoluminescence emission (PL) measurements the samples are excited by light of a fixed wavelength and the emission spectrum is recorded. The complementary method is to measure photoluminescence excitation (PLE) spectra, where the PL intensity is detected at a fixed emission wavelength and the excitation energy is varied; thus these spectra show the dependence of a specific emission wavelength on the energy of the excitation. In order to measure over a wide temperature range four different setups were used.

A commercial luminescence spectrometer (Perkin Elmer LS50B) allows PL and PLE spectra to be measured at room temperature under ambient conditions in reflection geometry. An excitation wavelength of 350nm was used for the PL spectra.

Measurements at 1.3K and at 6K were performed at the University of Stuttgart as follows. For measurements at 1.3K the samples were placed in a glass cryostat with superfluid helium. The samples were excited by the 363.8nm line of an argon ion laser (Spectra Physics) at a power of 10mW. The optical emission spectra were recorded with a system of filter, monochromator (Jobin-Yvon) and photomultiplier in a 90° setup with respect to the optical excitation path. For the measurements at 6K the samples were placed in a helium flow cryostat and were excited by the light of an XBO lamp (700W) passed through a H_2O filter and a double-monochromator (Spex). The emission of the samples was recorded in a 90° setup with a system of filter, double-monochromator (Spex) and photomultiplier. The excitation and emission spectra were corrected by the spectral characteristics of the XBO lamp, the monochromator used and the detector system.

Most of the measurements were made in our lab with the setup for a temperature range from 6K to 300K. The measurements were carried out in an optical continuous flow cryostat (Oxford CF200), where the samples were excited by the 441.6nm line (70mW) and/or the 325nm line (<10mW) of a He-Cd laser (Kimmon IK). The spectra were recorded either with a ¼m monochromator (CVI Digikröm DK240) coupled with a photomultiplier or photodiode, or with a spectrograph (SOLAR S-380, gratings: 150 lines/mm, 200 lines/mm, and 300 lines/mm) coupled with a liquid nitrogen cooled CCD detector (Princeton Instruments). All spectra were corrected by the specific spectral characteristic of the respective detection system.

For experimental reasons the luminescence spectra were recorded by measuring the intensity in intervals of the wavelength $d\lambda$. In order to plot the spectra over a linear scale of the corresponding energy (in eV or cm^{-1}) in intervals of $d\nu$, the spectra have to be corrected. From the relation $\nu=\lambda^{-1}$ directly follows

$$d\nu = \frac{d\nu}{d\lambda} d\lambda = -\frac{d\lambda}{\lambda^2}$$

Therefore, to plot the spectra over a linear energy-axis, the measured spectra have to be multiplied by λ^2 as was done for all spectra shown.

The CIE color coordinates of the PL spectra were measured with a Minolta Chroma Meter CS-100.

Transient and Delayed Luminescence

For measurements of the transient and delayed luminescence choppers in the optical path were added to the setups for the PL measurements described above. Transient signals were recorded with a monochromator coupled with a photomultiplier and a digital oscilloscope (HP 54504A). In order to separate the delayed part of the emission spectra from the prompt emission (described in Chapter 5.2), two synchronized light choppers were used. For measurements at temperatures $\leq$ 6K this was performed by two phase-stable 50Hz-choppers and in the setup for temperatures between 6K and 300K a Master-Slave chopper-system (HMS 220/221) was used.

UV-Vis-Spectroscopy

Absorption spectra of evaporated Alq_3 films with thicknesses between 10nm and 500nm were recorded with a commercial UV-Vis spectrometer (Perkin Elmer Lambda 2).

Optically Detected Magnetic Resonance (ODMR)

In the triplet state of a molecule the spins of two electrons are coupled to a state with the spin quantum number S=1. Due to the dipole-dipole interaction of the electron spins the energy of the triplet state splits into three sublevels without any external field. The zero field splitting tensor, which describes this interaction, is obtained by quantum mechanical calculations [McG69]. Using a system of principal axes, the tensor is completely determined by the zero field splitting parameters D and E, and the energy eigenvalues of the three sublevels can be obtained. From this the energetic splitting of the triplet state is $|D|+|E|$, $|D|-|E|$, and $2|E|$, respectively, as shown in Figure 3.2. With g as the Landé-factor and the Bohr magneton μ_B, the zero field splitting parameters are given by

$$D = \frac{g^2\mu_B^2}{2}\frac{3}{2}\frac{\langle\phi|r^2-3z^2|\phi\rangle}{r^5} \quad \text{and} \quad E = \frac{g^2\mu_B^2}{2}\frac{3}{2}\frac{\langle\phi|y^2-x^2|\phi\rangle}{r^5}.$$

D and E give information about the average quadratic distance of the electrons that are described by the orbital part of the wavefunction $|\phi\rangle$ in the corresponding coordinate system. Depending on the symmetry of the molecule the principal axes can be chosen to be coincident with the symmetry axes of the molecule.

The fine structure of the triplet state can be investigated by optically detected magnetic resonance (ODMR) at zero field. In the stationary regime under optical excitation there is a dynamic equilibrium between population and depopulation of the triplet system of the molecules. The population rates p_i of the sublevels $|T_i\rangle$ as well as their depopulation rates k_i are spin-sensitive and are usually different (i= x, y, or z). Therefore the population of the sublevels is different in the stationary regime, as indicated by the size of the circles in Figure 3.2. Transitions between two sublevels can be induced by using microwaves with the energy of

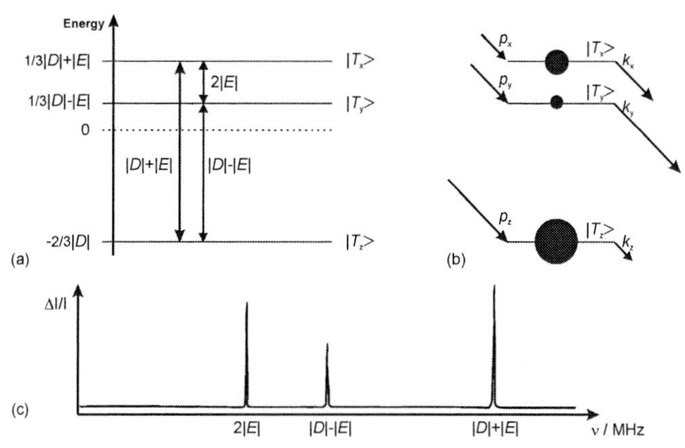

Figure 3.2: *(a): Schematic diagram of the zero field splitting of the triplet state.* $|T_x\rangle$, $|T_y\rangle$, *and* $|T_z\rangle$ *are the zero field spin levels with the eigenvalues* $1/3|D|+|E|$, $1/3|D|-|E|$, *and* $-2/3|D|$, *respectively. (b): Population and depopulation of the triplet state in the stationary regime.* p_i *and* k_i *are the population and depopulation rates, respectively, and the circles symbolize the number of molecules in the respective triplet state. (c): Schematic example of an ODMR measurement. At microwave frequencies corresponding to energy differences of the triplet sublevels a change in the PL-intensity I is observed. The relative intensities shown in this schematic figure are of no specific meaning.*

$|D|+|E|$, $|D|-|E|$, or $2|E|$. This changes the distribution between these sublevels and therefore the average lifetime of the triplet state. Consequently there is a decrease or increase of the amount of triplet states in the measured sample. In the first approximation molecules in the triplet state do not contribute to the fluorescence signal and therefore an increasing population of the triplet states results in a reduced PL intensity and vice versa. Thus changes in the triplet population due to resonant microwaves with energies of $|D|+|E|$, $|D|-|E|$, and $2|E|$ can be measured by changes in the PL intensity (see Figure 3.2 (c)). In our measurements on the Alq_3 samples the PL intensity was detected at the respective wavelength of the maximum of their PL spectrum. Further details on the principles of ODMR measurements are found in the literature [McG69, Cla82] and a detailed description of the setup used at the Institute of Physics III at the University of Stuttgart can be found in the PhD theses of Markus Braun and Oliver Wendland [Bra99, Wen03].

4 Thermal and Structural Properties of Alq_3 and Preparation of its Polycrystalline Phases

4.1 Polycrystalline Phases of Alq_3

This section describes the preparation and identification of polycrystalline phases of Alq_3 obtained by sublimation. To induce growth of different phases the temperature gradient in a sublimation tube was used. Phases that grow at different temperatures were obtained and then their crystalline structures were determined. Our results were compared with and completed by data meanwhile published by other researchers. A surprising outcome of this work was the discovery of a new blue luminescent crystalline phase of Alq_3, the δ-phase, which has significantly different properties compared to all other phases.

The temperature gradient sublimation described in Chapter 3 is a common

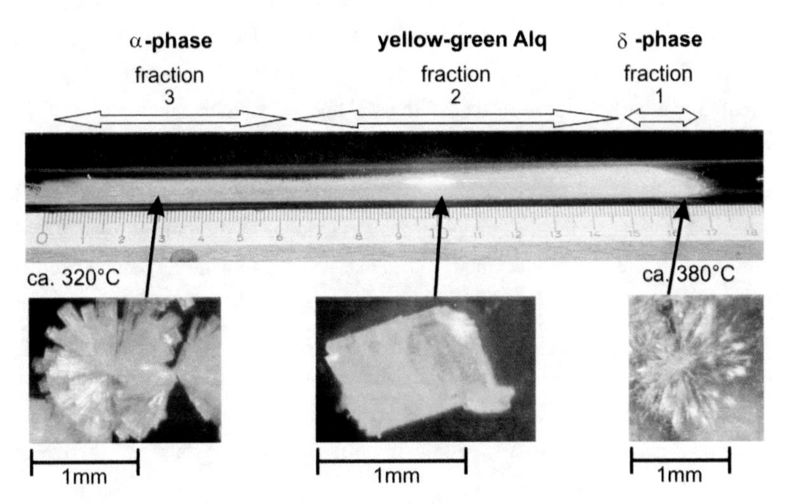

Figure 4.1: *Picture of a sublimation tube. Due to the temperature gradient in the sublimation tube, the material obtained is separated into three zones, which are labeled by fraction 1, fraction 2 and fraction 3. Crystals of these fractions in the tube are also shown.*

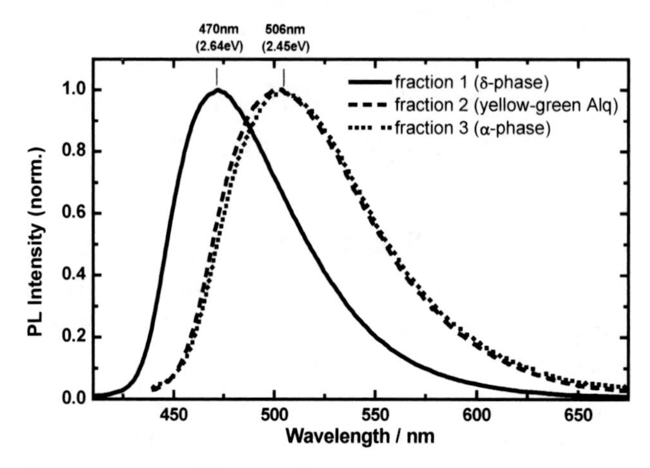

Figure 4.2: *PL spectra of the three fractions obtained from the sublimation tube, excited at 350nm and measured at room temperature*

method for purification of organic materials and was introduced in our lab for this purpose. After this purification procedure polycrystalline powders of different shapes were found in the sublimation tube and thus we distinguished between three different zones in the glass tube. The materials in these zones, hereafter called fractions, differ in their shape of crystals, their color, their solubility and their fluorescence.

A typical example of these glass tubes after sublimation is shown in Figure 4.1 with indicated areas for the three different fractions. In the hottest zone of the growth area there is an approximately 1.5cm wide region with very small needle-like crystals with white or slightly yellow appearance (fraction1). This zone is followed by the main fraction (about 8.5cm) with yellow cubic crystals and dimensions up to 500x500x500μm^3 showing yellowish-green fluorescence (fraction2). In the subsequent colder zone of the sublimation tube another fraction is obtained with dark yellow-green needle-like crystals with a size of 50x50x500μm^3 (fraction 3).

These fractions have different solubility in organic solvents. While fraction 3 and (apart from a small residue) also fraction 2 are readily dissolved in chloroform at relatively high concentration of more than 1% by weight, the solubility of fraction 1 is extremely poor. It takes several hours to dissolve a sizeable amount

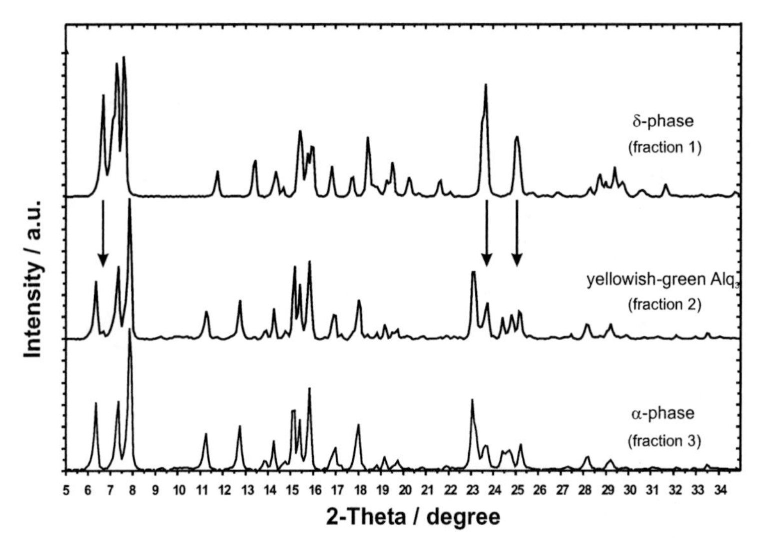

Figure 4.3: *X-ray powder diffractograms of polycrystalline Alq_3 fractions 1, 2, and 3 obtained from the sublimation tube in the 2Θ-range from 5 to 35 degrees (step width $\Delta 2\Theta=0.083°$). Arrows mark areas with the most significant differences.*

in chloroform, but then the color of the solution becomes similar to that of the other fractions.

Further differences between the three fractions are found in their PL spectra. Figure 4.2 shows these spectra with samples excited at 350nm and measured at room temperature. All fractions show one broad PL band with no additional structures and a tail at the side of longer wavelengths. Their main difference is the large blue shift of the PL maximum of about 0.19eV (36nm) from fraction 3 to fraction 1 with a PL maximum at about 506nm (2.45eV) and 470nm (2.64eV), respectively. The large blue shift is associated with a slight change in the shape of the PL spectrum, which is less symmetric for the blue luminescent fraction 1. The rise at the high energy side becomes steeper and the tail at low energies is more pronounced.

In order to investigate the cause of these differences, the crystal structures of the three fractions were determined by using X-ray powder diffraction spectroscopy. As a result of this data in principle two different crystal structures were found.

The crystal structures were determined from the X-ray spectra shown in Figure 4.3. Fraction 1 and fraction 3 show the main differences. These differences are best seen for small angles below 9 degrees and in the region between 22 and 26 degrees. From these two spectra the unit cells for fraction 1 and fraction 3 were determined[2]. Indexing of the observed peaks is given in tables in the appendix and the cell parameters determined for the different phases of Alq_3 are summarized in Table 4.1. The spectrum of fraction 2 seems to be a mixture of these two phases. Basically the spectrum is similar to that of fraction 3 apart from some very small peaks or shoulders. These additional peaks and shoulders are at positions where fraction 1 and fraction 3 are different, for example at 23.5 degrees and especially at 6.69 degrees. This suggests that fraction 2 mainly consists of the same phase as fraction 3, but has some small admixtures of material from fraction 1. The result that fraction 2 is a mixture of two different phases is relevant for applications, as it is mainly this fraction that is used for fabrication of OLEDs. From this X-ray data it becomes clear that the main difference is between fraction 1 and fraction 3, which have different crystal structures given in Table 4.1.

It is possible to compare these crystal data obtained above with results of other researchers, because a research group from Bologna in Italy started to investigate polycrystalline phases of Alq_3 independently. Their published data is also included in Table 4.1. They reported on three different crystalline structures called α-, β- and γ-phase [Bri00]. The published data for the α-phase are identical to those of fraction 3. β-Alq_3 is grown from solution and its properties are in principle similar to the α-phase, only with a small red shift in the PL due to slightly different intermolecular interaction in the crystal. The published data of γ-Alq_3 are listed in Table 4.1 for completeness, although they have to be considered as preliminary data and refined or improved data will be published in the near future.[3] In the paper by Brinkman et al. α- and β- Alq_3 are analyzed by X-ray diffraction and structural refinement, and the space group P-1 is

---

[2] In the first report on blue luminescent Alq_3 (fraction 1) in Reference [Bra01] we included the shoulder at 7.05° in the indexing, which leads to a different unit cell. The origin of this shoulder is discussed in detail in the following section.

[3] Comment of Dr. Muccini at the 290th WE Heraeus Seminar that an article on γ-Alq_3 had been recently submitted.

Table 4.1: *Crystallographic data of the polycrystalline phases of* Alq_3.

	α-phase (fraction 3) [Bri00, Bra01]	**β-phase [Bri00]**	**γ-phase [Bri00]**	**δ-phase (fraction 1) [Cöl02, Cöl03]**
crystal system	triclinic	triclinic	trigonal	triclinic
space group	P-1	P-1	P-31c	P-1
Z	2	2	2	2
a [Å]	12.91	10.25	14.41	13.24
b [Å]	14.74	13.17	14.41	14.43
c [Å]	6.26	8.44	6.22	6.18
α [°]	89.7	97.1	90	88.55
β [°]	97.7	89.7	90	95.9
γ [°]	109.7	108.6	120	113.9
V [Å^3]	1111	1072	1118	1072.5

determined for both phases [Bri00]. All phases and evaporated films are identified as consisting of the meridional isomer, and therefore only the meridional molecule was found at that time.

The denotation of the phases is in accordance with these published data. Fraction 3 and the main part of fraction 2 consist of the α-phase. The structure of fraction 1 is new and no corresponding phase has been published so far. Accordingly fraction 1 is hereafter called the δ-phase of Alq_3.

In this section two different crystalline phases of Alq_3 have been identified, namely the α- and δ-phase. δ-Alq_3 exhibits major differences to all other phases. It is a whitish powder, has a different crystal structure and, importantly, a strongly blue-shifted PL. On the other hand the α- and β- phase are very similar, as reported by Brinkmann et al. Consequently it is most interesting to investigate the differences and similarities of the α- and δ- phase of Alq_3, as will be done in the next sections.

4.2 Thermal Properties of Alq_3 and Controlled Preparation of the Blue Luminescent δ-Phase

As discussed in the previous section, the phases grow in different areas of the sublimation tube in regions of different temperature. Obviously, temperature has a strong influence on the formation of the different phases and thus it is important to learn more about the thermal properties of Alq_3. Therefore the formation conditions of the different phases of Alq_3 were investigated using differential scanning calorimetry (DSC) measurements in combination with structural and optical characterization.

Results

Figure 4.4 shows the DSC measurement of polycrystalline Alq_3 powder (α-phase) taken at a heating rate of 20°C/min. Coupled endothermic and exothermic peaks are observed at about 395°C prior to the large melting transition at 419°C. This additional phase transition has also been reported in the literature and has been attributed to polymorphism of the crystalline material [Bri00, Sap01]. It is very pronounced at fast heating rates (above 15°C/min). For slow heating rates the endothermic and exothermic transitions become broader, which is associated with a less intense peak compared to the strong melting peak. The peaks start to intermingle and are shifted to a slightly lower temperature, as shown in the inset of Figure 4.4 for heating rates of 20°, 10°, 5° and 2°C per minute. This behavior is similar to known irreversible monotropic solid-solid transitions [Met00, Wun90, For89]. Typically, the monotropic transition is slow and is mostly observed a few degrees below the melting point. Thus it is advisable to measure the monotropic transition isothermally at very slow heating rates. At higher heating rates (here above 2°C/min) it is easy to overrun the slow transition and thus reach the melting temperature of the metastable low-temperature phase, which results in the endothermic melting peak that merges directly into the exothermic crystallization peak. Although this process was not investigated in detail, the scenario described may explain the behavior shown in Figure 4.4.

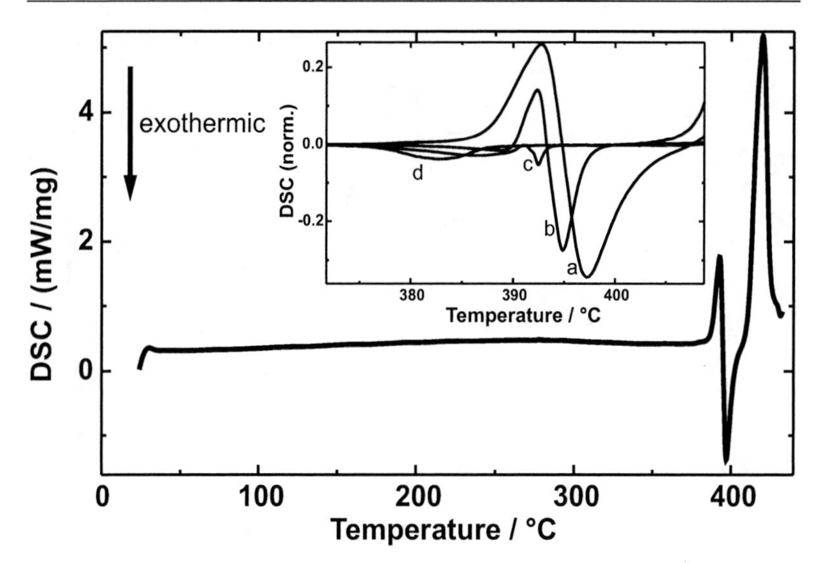

Figure 4.4: *DSC trace of Alq_3 with pronounced thermal transitions at 393°, 396° and 419°C measured at a heating rate of 20°C/min. Inset: Broadening and intermingling of the endothermic and exothermic peaks around 395°C in the DSC signal related to the sweep speed (a: 20°C/min, b: 10°C/min, c: 5°C/min, d: 2°C/min; normalized on the melting peak intensity). At low measuring speed only the more pronounced exothermic transition is visible.*

It should be noted that increasing the temperature above 430°C results in decomposition of the material and that a small broad transition at 320°C reported by Sapochak et al. [Sap01] was not observed in our samples. In the following measurements a slow heating rate of 2°C/min was used, where the shift of the peak temperatures is fairly small (see Figure 4.5) and where it is possible to stop the process at a defined temperature. Using this procedure the conditions for the preparation of different Alq_3 phases by a controlled thermal annealing process were specified.

In these slow DSC measurements three different regions are distinguished in Figure 4.5: In the first region (A) below the exothermic phase transition Alq_3 is the usual yellowish green powder, in the second region (B) between this phase transition and the melting peak Alq_3 is a whitish powder, and finally in region C

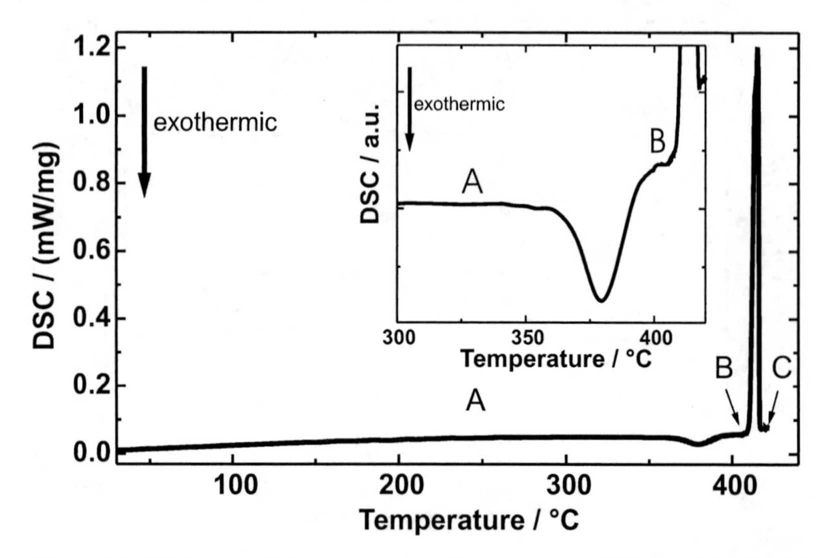

Figure 4.5: *DSC trace of Alq_3 measured at a heating rate of 2°C/min. The clearly pronounced exothermic phase transition at 380°C prior to the melting point is enlarged in the inset, as it becomes broad and less intense compared to the melting peak for this slow heating rate. A, B and C mark the regions of yellowish-green Alq_3, blue Alq_3 and melt, respectively.*

Alq_3 is a liquid melt. The glassy state of Alq_3 was obtained by quenching this melt in liquid nitrogen. Its highly amorphous character was verified by using X-ray powder diffraction measurements and using an image plate detection system. Cooling down the liquid melt slowly resulted in yellowish-green powder (A) again, as was previously reported [Nai93]. All of these materials are stable at room temperature.

Figure 4.6 shows the PL spectra measured at room temperature of annealed polycrystalline Alq_3 powder from regions A and B as well as of the quenched amorphous melt (C). For annealing temperatures up to 365°C Alq_3 is a yellowish-green powder with a PL maximum at 506nm (curve A). After the exothermic transition at about 380°C, there is a big blue shift of 0.18eV (37nm), associated with a slight change in the shape of the PL spectrum (curve B), which is less symmetric for the blue Alq_3. The quenched melt (curve C) is clearly red-

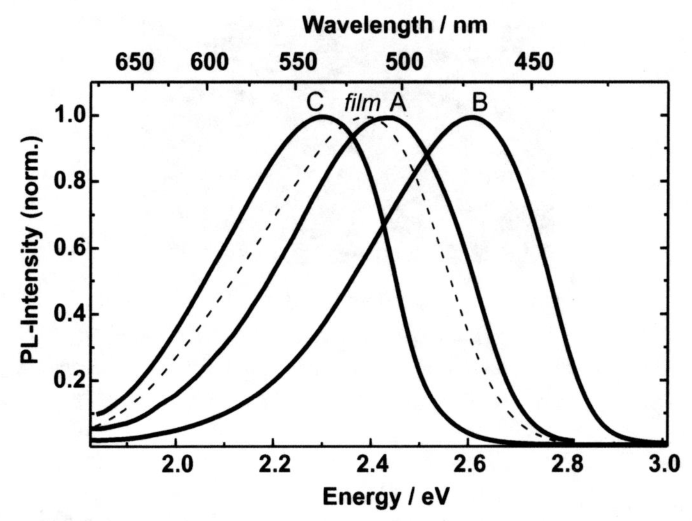

Figure 4.6: *PL spectra of Alq_3 samples taken from regions A, B and C of Figure 4.5, respectively, excited at 350nm. The PL of an evaporated Alq_3 film (dashed line) is shown for purposes of comparison. All spectra were measured at room temperature.*

shifted (0.14eV) compared to the yellowish-green Alq_3-powder (curve A). The strong difference in the emission color can be seen in Figure 4.7, where samples of the quenched melt, yellowish-green and blue Alq_3 are shown in daylight (a) and under UV-irradiation (b), respectively. The emission color is shifted from green (CIE coordinates: x=0.27, y=0.5) to blue (x=0.16, y=0.26). From Figure 4.7 one can also see the relatively low PL intensity of the quenched melt compared to the very intense PL emission of blue Alq_3. Therefore the PL quantum efficiency was measured and as a result the PL-QE of the quenched melt was found to be strongly reduced: The values obtained for blue Alq_3, yellowish-green powder, evaporated film and quenched amorphous melt were 51%, 40%, 19% and 3%, respectively. On the one hand this is the first report on blue luminescent Alq_3 obtained by a simple annealing process, and on the other hand the first report on such a pronounced red shift for amorphous Alq_3.

The dashed line in Figure 4.6 is the PL spectrum of an evaporated Alq_3-film as used in OLEDs. Although these films are commonly called "amorphous", one can clearly see that the PL maximum is located between the quenched melt and

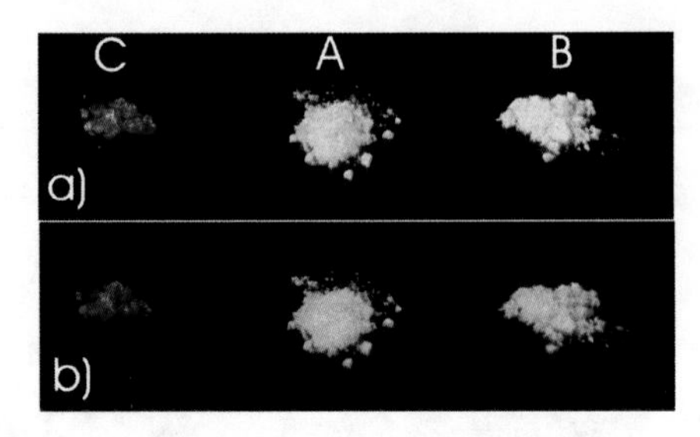

Figure 4.7: *Photographs of Alq_3 samples taken from regions A, B and C in Figure 4.5: a) in usual daylight and b) under UV-irradiation (excitation wavelength: 366nm), clearly showing the strong blue shift of the luminescence of the annealed material.(CIE color coordinates for A: x=0.27, y=0.50; for B: x=0.16 , y=0.26).*

crystalline Alq_3. This is an indication of the nanocrystalline character of these films, as reported by Tang et al. [Tan89]. The observed red shift of the quenched melt as compared to the crystalline material seems to be related to the morphology and intermolecular interaction. This effect is still under investigation and is beyond the scope of this work.

Yellowish-green Alq_3, blue Alq_3 and amorphous melt can be converted into each other. As described above, yellowish-green Alq_3 annealed above the phase transition results in blue Alq_3. Annealing blue Alq_3 above the melting point and cooling it down slowly, as shown in Figure 4.8ii), produced yellowish-green powder again. A pronounced recrystallization peak was observed at 326°C, similar to the results reported by Sapochak et al. [Sap01]. With the same procedure of annealing the quenched melt above the melting point and cooling it down slowly, yellowish-green powder is obtained again, and the quenched melt is converted into blue Alq_3 by annealing it between 380°C and 410°C. The successful conversion from one phase into the other was confirmed by measurements of the PL spectra, FT-IR spectra, Raman spectra, and X-ray diffraction.

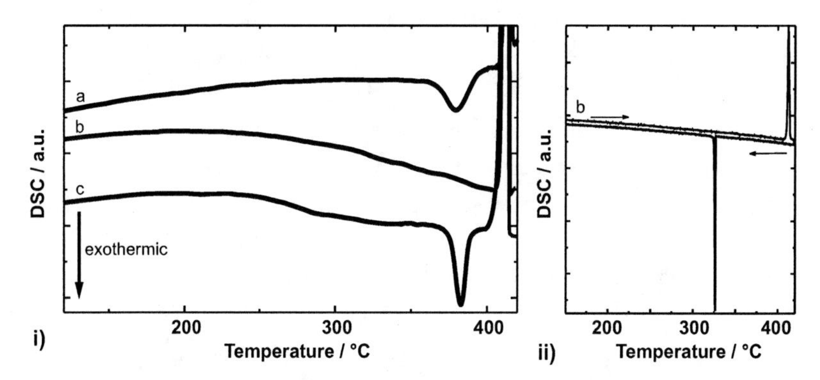

Figure 4.8: *DSC traces of a: yellowish-green Alq_3 and b: blue Alq_3. Trace c shows a second heating cycle after cooling down the melt (b) again. By annealing blue Alq_3 no phase transition at 380°C is observed (trace b in i) and ii)). Cooling down the melt gives a strong recrystallization peak at 326°C.*

Obviously, blue Alq_3 is formed during the phase transition at about 380°C. This phase transition appeared when starting the measurement with yellowish-green Alq_3, as shown in Figure 4.8. On the other hand, when starting the annealing procedure with blue Alq_3 material no such phase transition was observed, as shown in Figure 4.8i) and ii) trace b. However, measurements taken after the sample in Figure 4.8b had cooled down showed the exothermic peak again, as can be seen in trace c of Figure 4.8i).

As blue Alq_3 is formed in the region between the phase transition and the melting transition, the influence of temperature and preparation conditions in the region between 385°C and 410°C was investigated. Figure 4.9 shows X-ray powder diffraction spectra of blue Alq_3 prepared under three different conditions. For spectrum (I) yellowish-green Alq_3 powder (α-Alq_3) was annealed at 400°C for 2 hours. This spectrum is similar to the one obtained for fraction 1 in the sublimation tube shown in Figure 4.3. The shoulder at 2θ=7.05° for different samples of blue Alq_3 was variably pronounced. From this one may assume another high-temperature phase to be present in these samples. To test this, Alq_3 was annealed for several minutes at a higher temperature of 410°C (very close to the melting point) and a dark yellow substance was obtained, which exhibited only poor photoluminescence together with blue luminescent material. Its X-ray spectrum (Figure 4.9(II)) has a number of new peaks, which become very obvious for example at $2\Theta=7.05°$ (the position of the shoulder in spectrum (I)) and 25.85°. On the other hand, spectrum (III) shows Alq_3-powder annealed at 390°C for 6 hours. The additional lines observed in spectrum (II) are no longer present in this spectrum. Under these preparation conditions the second phase was not detected any more and one pure phase was obtained. The spectra are in a linear scale; thus "pure" in this connection means a weight proportion of 95% or more. The phase was determined to be the δ-phase of Alq_3. Indexing and unit cell parameters of this phase are given in the appendix and in Table 4.1.

Discussion

Blue luminescent Alq_3 obtained by annealing yellowish-green Alq_3 (α-phase) above the phase transition at about 380°C was identified as the δ-phase of Alq_3 with the unit cell given in Table 4.1. As seen in curves (I) and (II) of Figure 4.9, annealing Alq_3 at temperatures higher than 380°, close to the melting point, results in the appearance of new peaks in the X-ray spectra which can be attributed to an additional high temperature phase. Brinkmann et al. have already reported on such a high temperature phase, namely γ-Alq_3 [Bri00]. Using the given unit cell parameters from their work, the positions of all possible X-ray peaks for this phase were calculated as indicated by the vertical bars in curve (II) of Figure 4.9. These calculated peaks are located at the positions where spectrum (II) and (III) are different. Therefore it suggests that in sample (II) there is a high concentration of γ-Alq_3, whereas sample (III) is practically pure δ-Alq_3, as will be confirmed in the next section. From this it can be concluded that there are two high temperature phases of Alq_3: δ-Alq_3 and probably the γ-phase.

Blue luminescent Alq_3 obtained by train sublimation and by annealing showed the same behavior in respect of its solubility as well as its properties in PL, DSC, and IR measurements, confirming that in both cases the δ-phase of Alq_3 was obtained. In the sublimation tube the different phases were separated due to the temperature gradient. Since δ-Alq_3 and the other high temperature phase (most likely γ-Alq_3) are formed in a relatively narrow temperature region, the separation of the two phases by train sublimation is difficult and a certain ratio of

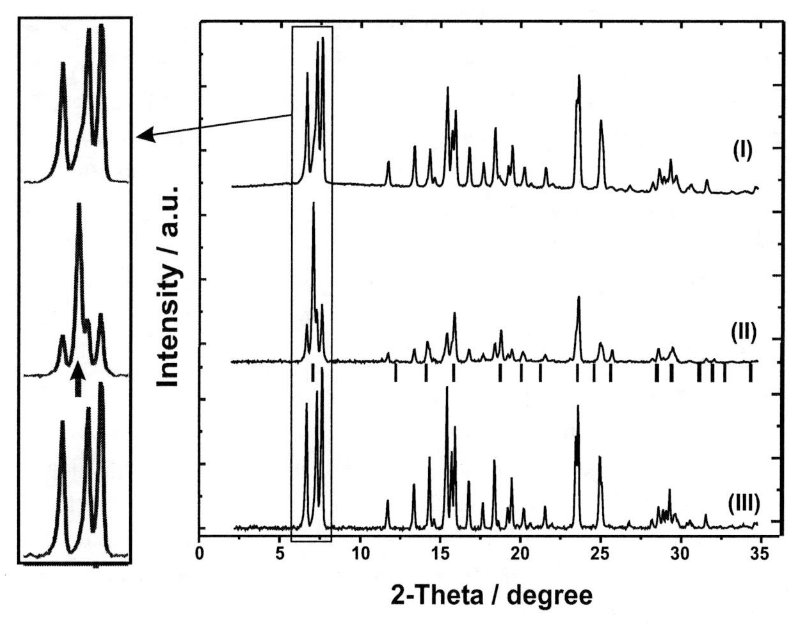

Figure 4.9: *X-ray powder diffractograms of polycrystalline blue Alq_3 prepared under different conditions. For spectrum (I) yellowish-green Alq_3-powder (α-Alq_3) was annealed at 400°C for 2 hours. In spectrum (II) the powder was annealed at 410°C (close to melting point). For spectrum (III) Alq_3 was annealed at 390°C for 6 hours. The additional lines and shoulders observed in spectrum (II) are not present in spectrum (III). Bars in spectrum (II) mark calculated positions for all possible X-ray peaks of γ-Alq_3.*

γ-Alq_3 is still present in the samples of δ-Alq_3, as indicated by the small shoulder at $2\theta=7.05°$ in the X-ray spectrum. On the other hand, under appropriate annealing conditions it is possible to obtain pure δ-phase without any visible admixtures of other phases, as demonstrated in curve (III) of Figure 4.9. A further advantage of this simple annealing process compared to temperature gradient sublimation is the possibility of obtaining large amounts (several grams) of pure δ-Alq_3 in a well-controlled process.

Chemical reactions during the annealing process can be excluded because the usual yellowish-green Alq_3 (α-phase) and the blue luminescent δ-Alq_3 can be easily converted into each other. Annealing yellowish-green Alq_3 at temperatures higher than 380°C results in δ-Alq_3, while heating δ-Alq_3 above the melting point and cooling the melt down slowly results in yellowish-green powder again. Another method of reconverting blue Alq_3 into yellowish-green Alq_3 is to evaporate the material or to dissolve it in any appropriate solvent (e.g. chloroform). The same holds for the glassy state of Alq_3 obtained by quenching the melt. It is readily dissolved in chloroform and films of good quality can be cast from such solutions. The PL spectrum of such films is the same as for evaporated films of Alq_3. By annealing material in the glassy state, it is possible to obtain both the yellowish-green α-Alq_3 and the blue δ-Alq_3, depending on the temperature. In all cases pure Alq_3 with no visible contaminating material is obtained. The possibility of transferring Alq_3 from one phase into the other implies that even at these high temperatures there is no decomposition or chemical reaction of the material. So it is important to emphasize that for all temperatures up to 425°C we are dealing with Alq_3, in agreement with ^{1}H NMR and FT-IR analysis of Alq_3 annealed at 422°C, where no decomposition products have been found [Sap01]. By excluding chemical reactions the difference in the phases must be of physical and structural origin.

The transition from yellowish-green Alq_3 into blue δ-Alq_3 by a simple annealing process can be used to obtain thin films of blue luminescent Alq_3. In preliminary experiments evaporated amorphous thin films with thicknesses from 300nm to 15μm, which were encapsulated (e.g. between two glass plates) to prevent the Alq_3 from volatilizing, were converted at 390°C into thin films showing blue luminescence. In addition it was possible to evaporate blue luminescent thin films directly onto heated glass substrates. These films of several micrometers are still comparatively thick, have polycrystalline structure, and have to be optimized as homogenous thin films with a thickness of several hundred nanometers are used in devices. This would enable OLEDs to be manufactured with blue luminescent δ-Alq_3. Further work to characterize such films and their application in OLEDs is in progress [Gär03].

In summary the formation conditions of different phases of Alq_3 were investigated using thermal, structural and optical measurements. δ-Alq_3 and an additional phase (probably γ-phase) were identified as high temperature phases of

Alq_3 and an efficient method of obtaining blue luminescent δ-Alq_3 by a simple annealing process is given. While the previously used train sublimation method resulted in only small amounts of δ-Alq_3 with admixtures of the γ-phase, it is now possible to prepare large amounts (several grams) of pure δ-Alq_3 by choosing appropriate annealing conditions. This is the prerequisite for further characterization of this blue luminescent phase and for the preparation of blue OLEDs from this material.

4.3 The Molecular Structure of δ-Alq_3

In the previous sections a new phase, the δ-phase, which exhibits major differences to all other phases, was introduced and characterized. It is also of interest to learn more about the structure of the molecule itself, and this issue shall be discussed in the following sections. This section describes high resolution X-ray measurements of the δ-phase identifying the facial isomer of Alq_3.

The problem in determining the structure of organic molecular crystals is mainly due to the large number of atoms (104 for Alq_3) in the unit cell. Standard laboratory equipment requires single crystals to solve the structure of a new phase of a material; however, so far single crystals large enough for a full analysis of the structure have only been available for the β-phase of Alq_3 [Bri00]. On the other hand, due to the use of high brilliance synchrotron radiation sources powder diffraction methods have progressed substantially in recent years, allowing very reliable determination of the structure from powder material without the need for larger single crystals. For this, high quality experimental data and specialized software for the analysis of the structure are required. These methods are very sensitive and unambiguous results are only expected if samples of one uniform crystal phase are measured. As the δ-phase can be isolated and δ-Alq_3 is easily obtained as a fine polycrystalline powder, these are good preconditions for this method. The measurements were performed by Dr. Robert Dinnebier (MPI Stuttgart) at the Synchrotron light source in Brookhaven, USA with an irradiation wavelength of λ=1.15Å.

In the case of a molecular crystal like Alq_3 it is necessary to start the simulation of the spectrum with a probable configuration of the molecules within the unit cell in order to achieve convergence within a reasonable calculation time. The start of the simulation was this assumed molecular configuration on the basis of the known connectivity of the molecule. The ligands were assumed to be planar and were randomly moved within a range of ±20° by a simulated annealing procedure until a minimized difference to the measured spectrum was obtained. After this, the position of the atoms was optimized by Rietveld refinements [Rie69]. The accuracy of the structure obtained is given by the R-values and the

Table 4.2: *Crystallographic data for δ-Alq_3. R_p, R_{wp}, and R-F^2 refer to the Rietveld criteria of the fit for profile and weighted profile respectively, defined by Langford and Louer [Lan96].*

Formula	$Al(C_9H_6NO)_3$	ρ-calc [g/cm^3]	1.423
Temperature [K]	295	2Θ range [°]	4-35.7
Formula weight [g/mol]	918.88	Step size [°2Θ]	0.005
Space group	P-1	Wavelength [Å]	1.14982(2)
Z	2	μ [1/cm]	2.48
a [Å]	13.2415(1)	Capillary diameter	0.7
b [Å]	14.4253(1)	R_p [%]	5.0
c [Å]	6.17727(5)	R_{wp} [%]	6.5
α [°]	88.5542(8)	R-F^2 [%]	10.5
β [°]	95.9258(7)	Reduced χ^2	1.6
γ [°]	113.9360(6)	No. of reflections	337
V [Å^3]	1072.52(2)	No. of variables	115

goodness of fit χ. In general three R values, R_p, R_{wp} and R-F^2, are given in the literature. The weighted profile factor R_{wp} and the Bragg R value R-F^2 are the most useful for following the progress of the refinement [Lan96, You82, Hil90]. The R_{wp} value is helpful when comparing subsequent refinements of the same data. When different analyses are compared the absolute value is less significant since the theoretical minimum value for R_{wp} is given by the expected factor R_{exp}, which is dependent on the quality of the data measured. For comparison of datasets the Bragg R value R-F^2 and χ should be used. Good values for R_p and R_{wp} are in the range of 5% [Mas00]. R-F^2 values below 15% are good and 10% shows a high accuracy of the refinement obtained with the real molecular structure [Mas00]. The ideal theoretical value for χ^2 is 1 and is never achieved due to experimental conditions. Acceptable values for χ^2 are below 2.25 [You93] and values between 1 and 2 are considered to be good. More information on the experimental procedure and analysis is found in Ref. [Cöl02] as well as in the literature [Cop92, You93, Lan96, Mas00].

The following analysis of the data of the δ-phase of Alq_3 was made on the assumption of the facial isomer as well as on the assumption of the meridional isomer being the constituent of this phase. First the results for the facial isomer are given, followed by the results for the meridional isomer for comparison. The data analysis assuming the facial isomer gave by far the best results.

Results

Figure 4.10 shows the observed spectrum together with the best Rietveld-fit profiles for the assumption of the facial isomer. The enlarged difference curve between observed and calculated profiles is given in an additional window below. Indexing of this very well resolved powder spectrum with the ITO routine [Vis69] led to a primitive triclinic unit cell for Alq_3 with lattice parameters given in Table 4.2. The number of formula units per unit cell could be determined as Z=2 from packing considerations and density measurements. P-1 was selected as the most probable space group, which was confirmed by Rietveld refinements. The high quality of the refinement becomes obvious from the excellent differential pattern in particular at high diffraction angles (corresponding to small distances in real space), the R_{wp} value of 6.5%, and the Bragg R value $R\text{-}F^2$ of 10.5%. Crystallographic data, positional parameters as well as selected bond lengths and angles for δ-Alq_3 are listed in Table 4.2 and in the corresponding figures and tables in the appendix.

The molecular structure of δ-Alq_3 obtained from these measurements is shown in Figure 4.12. Compared to the idealized isolated facial Alq_3 isomer, the molecule is only slightly distorted, which reduces its symmetry only negligibly, and the planes defined by the O- and N-atoms, respectively, are parallel. The molecules form linear stacks in the c-direction of the crystal. The projection along the c-axis as well as the projection perpendicular to the planes of the hydroxyquinoline ligands, which shows the overlap between ligands of neighboring Alq_3 molecules, are shown in Figure 4.13.

The data was also evaluated on the assumption of the meridional isomer. The best fit obtained for this case is plotted in Figure 4.11 together with the differential curve. A comparison with Figure 4.10 clearly shows that the fit assuming the meridional isomer is by far worse than the result for the facial

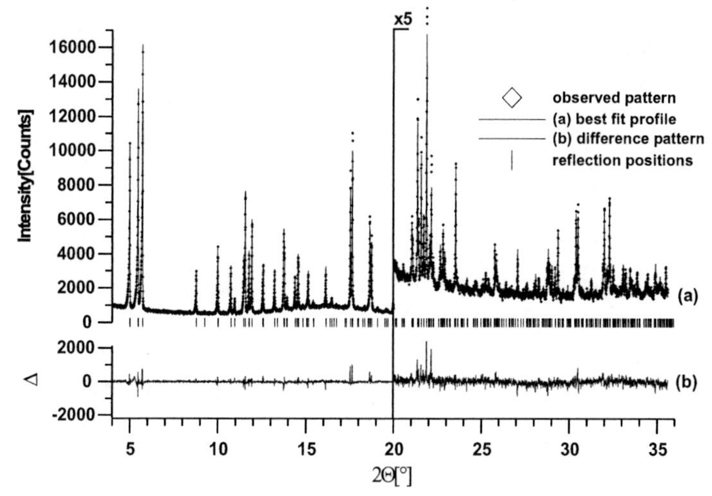

Figure 4.10: *Scattered X-ray intensity for δ-Alq_3 under ambient conditions as a function of diffraction angle 2Θ. Shown are the observed patterns (diamonds), the best Rietveld-fit profiles on the assumption of a facial isomer (line) and the enlarged difference curves between observed and calculated profiles in an additional window below. The high angle part is enlarged by a factor of 5, starting at 20°. The wavelength was λ = 1.15 Å.*

isomer. Refinement resulted in a distorted meridional molecule, whereby the distance for one coordination bond (Al-N) was elongated more than 10% compared to the others (ligand A and B: ca. 2.1Å, ligand C: 2.39Å) and a Bragg R value R-F^2 of 19.4% was obtained. R-Values, tables and a picture of the distorted meridional molecule are given in the appendix.

Discussion

The most important outcome of these refinements is that the δ-phase of Alq_3 consists of the facial isomer. For a long time it was believed that the facial isomer is unstable and would not exist. Thus, the results shown here are the first evidence for the existence of this facial isomer [Cöl03, Cöl03a]. The simulations assuming the facial isomer closely match the measured spectrum, as can be seen in the differential spectrum in Figure 4.10, which is much better than the differential spectrum in Figure 4.11 of the best possible fit for the meridional

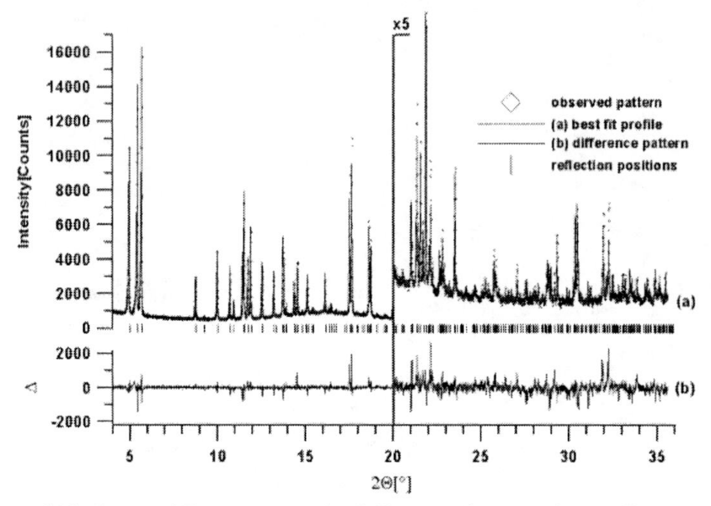

Figure 4.11: *Scattered X-ray intensity for δ-Alq_3 at ambient conditions. Shown are the observed patterns (diamonds), the best Rietveld-fit profiles on the assumption of a meridional isomer (line) and the enlarged difference curves between observed and calculated profiles in an additional window below. Best values obtained for R_p, R_{wp} and $R\text{-}F^2$ are 7.3%, 9.4% and 19.4%, respectively.*

isomer. For the meridional isomer the molecule is distorted and a substantially lower Bragg R value (by 9%) was obtained compared to the facial isomer ($R\text{-}F^2$= 10.5% facial, 19.4% meridional). The R values for the facial isomer indicate a high quality of the refinement, resulting in a very high probability that the δ-phase consists of this isomer. Furthermore, the high quality of the fit and the very well resolved spectrum suggests that the samples of δ-Alq_3 are an almost pure phase, confirming the results in Chapter 4.2. Therefore it can be concluded that the δ-Alq_3 samples prepared under defined annealing conditions as described above are a pure phase without significant admixtures of other phases and that δ-Alq_3 consists of the facial isomer. Thus, as a result of the preparation of δ-Alq_3, we have for the first time successfully isolated the long sought-after facial isomer of Alq_3.

The data also gives information about distance and orientation of the molecules and thus about molecular packing in the crystal. It is noteworthy that the molecules are arranged in a manner minimizing the possible overlap of the π-

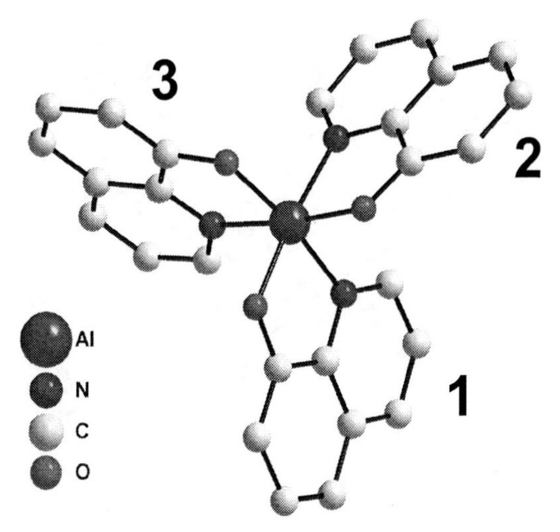

Figure 4.12: *Facial Alq_3 molecule of the δ-phase with the three hydroxyquinoline ligands labeled by 1, 2 and 3. H-atoms are omitted for simplicity.*

orbitals between pairs of hydroxyquinoline ligands belonging to neighboring Alq_3 molecules, as shown in Figure 4.13. As demonstrated by Brinkmann et al., the orbital overlap influences the optical properties and can explain shifts in the photoluminescence spectra of different phases of Alq_3 [Bri00]. In δ-Alq_3 the pyridine rings of antiparallel ligands 1 face each other with an interligand distance of 3.4Å (Figure 4.13(a)). The partial overlap of the rings is smaller as compared to the other known phases, and the atoms are slightly displaced, further reducing the overlap of the π-orbitals. Figure 4.13(b) and (c) show the projection perpendicular to the planes of Ligand 2 and Ligand 3, respectively. The interligand distance is about 3.45 Å and these ligands do not overlap at all. Thus a strongly reduced π-orbital overlap of neighboring ligands is found in δ-Alq_3 as compared to the α- and β-phase. As only one ligand of each molecule overlaps with a neighboring molecule, there are no π-π links generating an extended one-dimensional chain as reported for the β-phase [Bri00]. In view of this, both the packing effect with reduced intermolecular interaction and the changed symmetry of the molecule are likely to be responsible for the large blue-shift of the photoluminescence by 0.2eV, which is in the same range as predicted theoretically by Curioni et al. for the two isomers [Cur98].

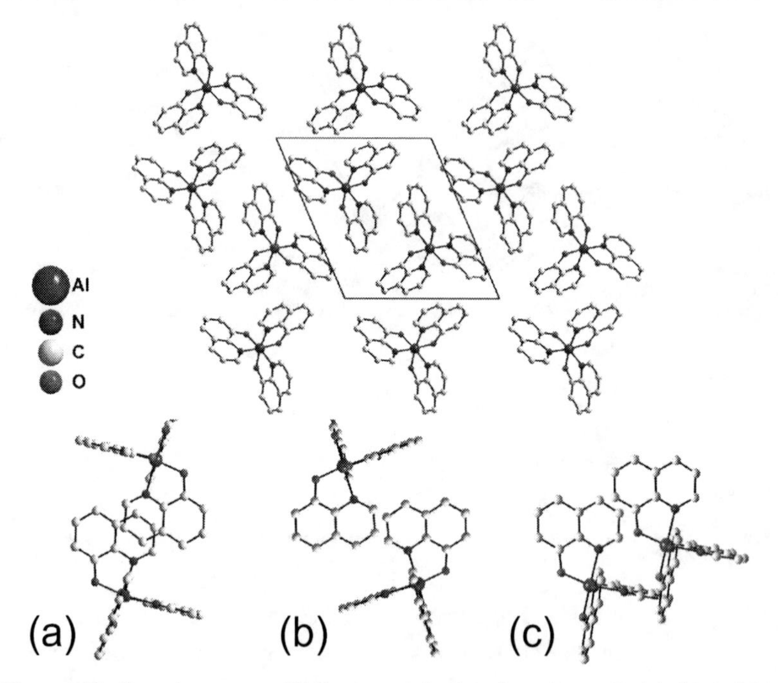

Figure 4.13: *Crystal structure of δ-Alq_3 in a projection along the c-axis. (a), (b), and (c) are projections perpendicular to the planes of the hydroxyquinoline ligands 1, 2, and 3, respectively, showing the overlap between ligands of neighboring Alq_3 molecules.*

For transformation from the meridional isomer to the facial isomer one ligand, namely ligand C in Figure 2.1, has to flip by 180°. From our results the facial isomer is formed at temperatures above 380°C; thus the question is of interest whether this transition is energetically allowed for this molecule. Amati et al. made theoretical calculations for several possible transition processes between the geometric isomers of Alq_3 and its stereoisomers and they found that the thermal conversion from the meridional isomer to the facial isomer is energetically possible [Ama02]. Very recently Utz et al. reported on NMR measurements of solutions elucidating an internally mobile nature of the Alq_3 complex [Utz03]. They found a high probability of ligands flipping by 180° and suggested that this process takes place on a time scale of about $5s^{-1}$ at room temperature in solution. In these measurements they were only able to determine the meridional isomer for two reasons: First, the facial isomer is predicted to be less stable by about 17kJ/mol for the isolated molecule [Cur98, Ama02], thereby reducing its lifetime in solution; second only the flip of ligand C may result in the facial isomer giving a lower probability for this process, and thus the expected concentration of this isomer in solution is likely to be too small to be measured [Utz03]. These measurements and the theoretical work of Amati et al. demonstrate that the transformation from the meridional isomer to the facial isomer at elevated temperature is possible as was carried out for the δ-phase.

In summary X-ray powder diffraction measurements were analyzed for the two possible geometric configurations of the Alq_3 molecule. By far the best results were obtained for the facial isomer, and the high quality of the fit gives convincing evidence that the δ-phase consists of the facial isomer. Very recently this was also confirmed by NMR measurements [Utz03a]. The molecular structure was determined and packing in the crystal with intermolecular interaction due to the overlap of the π-orbitals was discussed. Nevertheless, the conclusive evidence of its structure would be given by analysis of single crystals of δ-Alq_3. In fact, very recently the group at Eastman Kodak made structural analysis on single crystals of δ-Alq_3, prepared as described in this book, which confirmed our findings on the structure of the facial isomer and its packing in the crystal [Tan03]. Furthermore, the existence of the facial isomer also suggests differences in the molecular vibrational modes and differences in the excited states of δ-Alq_3 compared to the other phases, as will be discussed in the next chapters.

4.4 Vibrational Analysis

Due to the different molecular symmetry of the meridional and facial isomers (C_1 versus C_3), vibrational analysis using IR spectroscopy should be one possible way to differentiate between them. In particular, the first coordination sphere or central part of the molecule AlO_3N_3 should show characteristic vibrational properties for each isomer (Al-O and Al-N modes located below $600cm^{-1}$, as calculated by Kushto et al. [Kus00]). Furthermore, there is a weak coupling of the three ligands via the central part, and movements around the central aluminum atom are involved in most of the molecular vibrations below $1700cm^{-1}$. This coupling depends on the relative positions of the oxygen atoms of the ligands. For the facial isomer each oxygen atom faces a nitrogen atom, and thus the coupling via the Al atom is identical for all ligands, whereas for the meridional isomer one can clearly distinguish between the ligands labeled by A, B and C in Figure 2.1. For the meridional isomer, the coupling mainly affects the ligands B and C, where the oxygen atoms face each other, and to a lesser extent the A and B ligands, which have the oxygen and nitrogen atoms opposite. The coupling mechanism of ligand A and C is mainly characterized by the modes of the two opposite nitrogen atoms. This means that due to the lower symmetry of the meridional molecule each vibrational mode has a slightly different energy for the three ligands.

Degli Esposti et al. have published a detailed theoretical study of the vibrational properties of the meridional Alq_3-molecule [Esp02]. For the first time the coupling terms between the three ligands were explicitly taken into account. This interaction of the ligands, mainly via the Al-bonds, is expected to be different for the two isomers. An important result of their paper is a well-founded understanding of the IR spectra of meridional Alq_3 with the possibility of attributing lines to the three different ligands A, B and C as labeled in Figure 2.1. Based on this detailed theoretical work and calculations made for the two isomers by Kushto et al., first the experimental results on polycrystalline Alq_3 for the entire spectrum up to $1700cm^{-1}$ will be discussed. Thereafter emphasis will be laid on the stretching modes around the aluminum atom (below $600cm^{-1}$) before investigating the influence of crystallinity on the vibrational spectra.

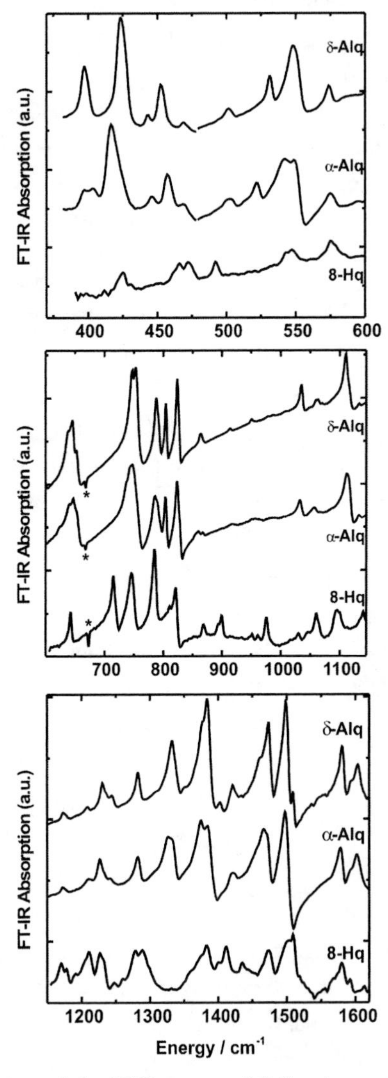

Figure 4.14: *Comparison of the FTIR-spectra of δ-Alq_3 (upper trace), α-Alq_3 (middle trace) and hydroxyquinoline (8-Hq, lower trace) in the range from 350cm^{-1} to 1650cm^{-1}. Below 420cm^{-1} the KBr optic reduces the quality of the spectrum and thus α-Alq_3 and δ-Alq_3 were measured with mylar-coated beamsplitters, CsI windows and samples on CsI pellets in the region between 370 and 480cm^{-1}. The peak marked with an asterisk (*) is an artifact of the setup and is not related to the sample.*

Table 4.3 summarizes the measured band positions and their relative intensities together with their preliminary assignments for the meridional Alq_3 molecule, which is the constituent of α-Alq_3 [Bri00, Kus00]. For the δ-phase of Alq_3 it was shown by structural investigations in Chapter 4.3 that it consists of the facial isomer. We can therefore use the IR spectra to identify characteristic differences in their vibrational properties [Cöl03a]. (Raman measurements lead to the same results and are therefore not discussed in this section).

As can be seen in Figure 4.14, most of the bands in δ-Alq_3 have slightly reduced line width. Furthermore, broad featureless bands in α-Alq_3 become narrower and exhibit pronounced structures in δ-Alq_3. In particular some adjacent bands in α-Alq_3 become one single peak in δ-Alq_3. Examples are found at wavenumbers $1470cm^{-1}$, $1380cm^{-1}$, $1333cm^{-1}$, $864cm^{-1}$, $750cm^{-1}$, and $638cm^{-1}$. According to Degli Esposti et al. the broad band of α-Alq_3 at $645cm^{-1}$, consisting of peaks at $652cm^{-1}$, $647cm^{-1}$, $640cm^{-1}$ and the small shoulder at about $625cm^{-1}$, is due to the pyramidalization mode of the nitrogen atoms within the different ligands A, B and C under participation of Al-O stretching. Whereas the modes at $652cm^{-1}$ to $640cm^{-1}$ are mainly the result of torsional and pyramidalization modes of ligand C, the shoulder at $625cm^{-1}$ originates mainly from ligand A, indicating the inequivalence among the ligands in the meridional isomer of Alq_3. In the facial isomer the ligands are equivalent and therefore the low frequency shoulder should not be observed in δ-Alq_3. It can also be noted that the whole band is narrower for the δ-phase with a full width at half maximum (FWHM) of $28cm^{-1}$ and $19cm^{-1}$ for α- and δ- Alq_3, respectively. For α-Alq_3, the band centered at $747cm^{-1}$ consists of two overlapping bands, one involving $C\text{-}C^{ABC}$ stretching under participation of Al-O stretching modes on the low frequency side and a second involving $C\text{-}C^{ABC}$ deformations on the high frequency side. Each of the bands consists of contributions from ligands A, B and C. In this band A-vibrations are found at the flanks and C-vibrations in the center. Contributions from B are in between the contributions of A and C. If one also takes into account a possible splitting due to the crystal field, it is clear why this band is so broad. In contrast, for δ-Alq_3 the FWHM again is smaller (α: $40cm^{-1}$, δ: $17cm^{-1}$) and two peaks are clearly resolved. For the facial isomer the ligands are equivalent and only one peak is expected; therefore this splitting can be attributed to a crystalline effect (as will be discussed in more detail below).

Weak but interesting bands are observed at $867cm^{-1}$ and $859cm^{-1}$ for α-Alq_3 and assigned to torsional and pyramidalization modes of the different ligands. The peak at $867cm^{-1}$ is due to vibrations of ligand B, while the peak at $859cm^{-1}$ originates from vibrations of ligands A and C. For δ-Alq_3, however, instead of this splitting only one single peak at $864cm^{-1}$ is observed. The difference between the ligands seems indeed to be removed in δ-Alq_3, as expected for the facial isomer. The superposition of two peaks at $1332cm^{-1}$ and $1327cm^{-1}$ results in a broad band for α-Alq_3. This band is a result of CCH^{ABC} bending modes of the different ligands and C-O^{ABC} stretching modes. Theoretically the C-O stretching bands of ligand A and C are separated by more than $6cm^{-1}$. The C-O stretching mode of ligand A is located at the high frequency side and the C-O stretching mode of ligand C at the low frequency side of this band. Consequently a broad band is observed in the IR-spectrum. Instead of this broad feature there is only one peak for δ-Alq_3 located at $1333cm^{-1}$. This is another clear indication of the equivalence of vibrational modes in the three ligands of the facial isomer in the δ-phase.

For the doubled band observed at about $1380cm^{-1}$ in α-Alq_3, splittings of 2-$3cm^{-1}$ are predicted due to C-C^{ABC} and C-N^{ABC} stretch vibrations in the different ligands, which are not resolvable. The clear doublets in both crystalline phases (α and δ) are thus attributed again to a crystalline effect, as will be discussed later. Another intense and broad feature is observed for α-Alq_3 at $1468cm^{-1}$, namely a superposition of C-O^{ABC} and C-C^{ABC} stretching vibrations in the ligands A, B and C of the meridional molecule. Theoretical energetic distances between these contributions are $4cm^{-1}$ and $8cm^{-1}$, resulting in one broad band. The weak low frequency shoulder is assigned to C-H bending modes in both phases, but for the main band only one peak is expected for the facial isomer, as is observed for δ-Alq_3 at $1473cm^{-1}$.

Table 4.3: *Comparison of the experimental frequencies of the δ-phase (E(δ)), the α-phase (E(α)), and 8-hydroxyquinoline (E_{quin}) together with the assignments according to Kushto et al. and Degli Esposti et. al. [Kus00, Esp02]. For the α-phase the assignments consider the contribution of the three different ligands in meridional Alq_3. A, B and C refer to the corresponding ligands and the theoretical energies for the meridional Alq_3 are listed under (E_{calc}). In the facial isomer the ligands cannot be distinguished; therefore the assignments are given separately for clarity and also to demonstrate the reduced number of vibrational modes in the facial isomer. In the last column the normal modes of 8-hydroxyquinoline are labeled by the symmetry within the C_S group. Relative intensities for the δ- and the α-phase are given in column I(δ) and I(α). Asterisks mark peaks related to crystallinity.*

No.	E (δ) [cm^{-1}]	I(δ) [a.u.]	Mode	E (α) [cm^{-1}]	I(α) [a.u.]	E_{calc} [cm^{-1}]	Mode	E_{quin} [cm^{-1}]	Symmetry
1	1604	57	CC-str., CCC-bend.	1603	59	1606.5	CC-str.^{A}, CC-str.^{B}		
						1605.4	CC-str.^{C}, CCC-bend.^{C}		
2	1594	23	CCC-bend., CC-str.	1594sh	27	1601.7	CCC-bend.^{B}, CC-str.^{C}		
								1592	9 A'
						1583.4	NC-str.^{A}, CC-str.^{A}	1580	10 A'
3	1580	89	NC-str., CC-str.	1578	72	1582.4	NC-str.^{C}, CC-str.^{C}		
						1575.7	NC-str.^{B}, CC-str.^{B}		
4	1509	25	11A'					1508	11 A'
						1503.2	CC-str.^{B}		
5	1498	183	CC-str., CCH-bend.	1497	150	1500.6	CC-str.^{C}, CCH-bend.^{C}	1503	
						1499.7	CC-str.^{C}, CCH-bend.^{B}		
6						1476.1	CO-str.^{A}, CC-str.^{A}		
7	1473	141	CO-str., CC-str.	1471	96	1468.0	CO-str.^{C}, CC-str.^{A}	1471	12 A'
8	1463	76	CC-str., CO-str.			1462.7	CC-str.^{B}, CO-str.^{B}	1471	12 A'
						1426.9	CCH-bend.^{A}	1434	13 A'
9	1421	39	CCH-bend.	1421	17	1426.2	CCH-bend.^{C}		
						1425.0	CCH-bend.^{B}		
								1410	14 A'
10	1403	11							
						1392.3	NC-str.^{B}, NCH-bend.^{B}		
11	1384	194	NC-str., NCH-bend.	1385	112	1391.4	NC-str.^{C}, NCH-bend.^{C}		
						1389.8	NC-str.^{A}, NCH-bend.^{A}		
						1376.7	CC-str.^{A}, CC-str.^{B}	1381	15 A'
12	1377	143	CC-str.	1375	117	1375.8	CC-str.^{C}, CC-str.^{A}		
						1372.9	CC-str.^{B}, CC-str.^{A}	1372sh	
						1337.2	CCH-bend.^{A}, CO-str.^{A}		
13	1333	97	CCH-bend., CO-str.	1327	86	1334.6	CCH-bend.^{B}, CCH-bend.^{C}	1355	16 A'
						1331.2	CCH-bend.^{C}, CO-str.^{C}		
								1285	17 A'
						1294.3	NC-str.^{A}, CCC-bend.^{A}		
14	1282	55	CO-str., NC-str. CCH-bend.	1282	48	1291.7	NC-str.^{B}, CO-str.^{B}	1276	18 A'
						1290.1	CO-str.^{C}, CCH-bend.^{C}		
15	1244	25	CCH-bend.	1242	10		CCH-bend.	1244	
						1229.1	NC-str.^{A}, NCH-bend.^{A}		
16	1230	52	NC-str., NCH-bend.	1227	54	1226.9	NCH-bend.^{C}, NC-str.^{C}		
						1221.6	NCH-bend.^{B}, NC-str.^{B}	1223	19 A'
						1216.9	CC-str.^{B}	1206	20 A'
17	1209	10	CC-str.	1211	16	1216.3	CC-str.^{C}		
						1216.2	CC-str.^{A}		
								1187	
						1169.3	CCH-bend.^{C}, CCH-bend.^{A}	1173	
18	1173	20	CCH-bend.	1173	13	1168.6	CCH-bend.^{B}	1166	21 A'
						1168.3	CCH-bend.^{A}		
						1134.2	CCH-bend.^{C}	1140	22 A'
19	1133	10	CCH-bend.	1133	3	1135.5	CCH-bend.^{A}		
						1133.1	CCH-bend.^{B}		
						1105.5	CCH-bend.^{A}, CNC-bend.^{A}	1113	
20	1111	143	CCH-bend., CNC-bend.	1112	120	1104.3	CCH-bend.^{C}, CNC-bend.^{C}	1099	23 A'
						1102.5	CCH-bend.^{B}, CNC-bend.^{B}		
						1056.1	CC-str.^{A}		
21	1060	22	CC-str.	1058	20	1052.0	CC-str.^{C}	1060	24 A'
						1050.1	CC-str.^{B}		
						1034.8	CC-str.^{A}		
22	1036	70	CC-str.	1033	47	1030.3	CC-str.^{C}	1029	
						1029.1	CC-str.^{B}		
								974	1 A''
								959	2 A''

No.	E (δ) [cm^{-1}]	I(δ) [a.u.]	Mode	E (α) [cm^{-1}]	I(α) [a.u.]	E_{calc} [cm^{-1}]	Mode	E_{quin} [cm^{-1}]	Symmetry
								950	3 A''
								897	26 A''
23				867	12	868.5	CC-tors.^{B}, C-pyr.^{B}		
24	864	31	CC-tors., C-pyr.	859	19	865.6	CC-tors.^{C}, C-pyr.^{C}	866	4 A''
						864.8	CC-tors.^{A}, C-pyr.^{A}		
						826.2	C-pyr.^{B}	818	5 A''
25	823	227	C-pyr.	823	197	824.3	C-pyr.^{C}		
						820.2	C-pyr.^{A}		
						793.2	CCC-bend.^{C}		
26	804	168	CCC-bend.	803	152	792.4	CCC-bend.^{B}	807	27 A''
						791.7	CCC-bend.^{A}		
						787.6	CC-tors.^{B}, C-pyr.^{B}		
27	788	182	N-pyr., CC-tors. C-pyr.	786	150	786.2	N-pyr.^{C}, CC-tors.^{C}	781	6 A''
						784.2	C-pyr.^{A}, CC-tors.^{A}		
						751.8	C-pyr.^{B}		
28	753*	250	C-pyr. *	747	250	750.3	C-pyr.^{C}		
						749.1	C-pyr.^{A}		
						746.0	OAl-str.^{A}, CC-str.^{A}		
29	747*	239	OAl-str., CC-str. CCC-bend.*	747		743.2	OAl-str.^{C}, CCC-bend.^{C}	741	7 A''
						741.0	OAl-str.^{B}, CC-str.^{B}	711	28 A''
30	652	98	N-pyr., CC-tors.	652	111	653.4	N-pyr.^{C}, CC-tors.^{C}		
						649.5	N-pyr.^{C}, CC-tors.^{C}		
31	645	179	N-pyr., C-pyr.	647	145	648.8	N-pyr.^{C}, C-pyr.^{A}		
32	638	150	N-pyr.	640	127	646.8	N-pyr.^{B}, N-pyr.^{C}	637	
33				625sh	60	639.8	N-pyr.^{A}, COAl-bend.^{A}		
						570.5	CCC-bend.^{C}, COAl-bend.^{C}		
34	574	56	COAl-bend, CCC-bend.	575	46	568.8	COAl-bend.^{A}, CCC-bend.^{A}	575	10 A''
						566.9	COAl-bend.^{B}, COAl-bend.^{C}		
35	548	174	OAl-str.	549	144	540.4	OAl-str.^{B}, OAl-str.^{C}	547	
36				542	141	535.7	OAl-str.^{A}, CCC-bend.^{A}	543	30 A''
37	531*	77	OAl-str., CCC-bend.*	522	54	512.0	OAl-str.^{B}, CCC-bend.^{B}		
						495.6	CCC-bend.^{B}, CCC-bend.^{C}		
38	501	31	CCC-bend.	503	33	493.0	CCC-bend.^{B}, CCC-bend.^{C}		
						490.4	CCC-bend.^{A}	491	31 A''
						476.0	C-pyr.^{B}, NAl-tors.^{C}		
39	472	21	NAl-tors., C-pyr., CC-tors., N-pyr.	471	59	474.6	NAl-tors.^{C}, N-pyr.^{C}	471	11 A''
						473.5	C-pyr.^{A}, CC-tors.^{A}		
40	469	23	N-pyr., NAl-tors.	465sh	35	464.3	N-pyr.^{C}, NAl-tors.^{C}	465	32 A'
41	452	96	N-pyr.	454	145	462.5	N-pyr.^{C}, N-pyr.^{A}		
42	443	26	N-pyr., NAl-tors.	442	69	450.3	N-pyr.^{C}, NAl-tors.^{C}		
43	423	260	Al-N-str.	418	206	433	Al-N-str.		
44				405	49	415	Al-N-str.		
45	397	117	Al-N-str.	396	45	407	Al-N-str.		

Further arguments for different isomers in α- and δ-Alq_3 arise from the symmetry of the first coordination sphere or central fragment around the Al-atom. As discussed above, the differences in the IR spectra of the two phases, in particular the higher number of IR-active vibrational modes for the α-phase in comparison to the δ-phase of Alq_3, can be ascribed to the different symmetry of the constituent meridional and facial isomers, namely C_1 and C_3, respectively. If we consider the central fragment AlO_3N_3, the local symmetry for each isomer is C_{2v} and C_{3v}, respectively, as shown in Figure 4.15. The separation of the central part from the ligands is justified by the different and well-separated vibrational energies belonging to these groups, as observed in the comparison of Alq_3 with the hydroxyquinoline parent of the ligands in Figure 4.14 [Lar68]. Thus IR-modes that belong to the AlO_3N_3 group are found predominantly in the region below 600cm^{-1}. Particular focus is on the stretching vibrations of this central part. For

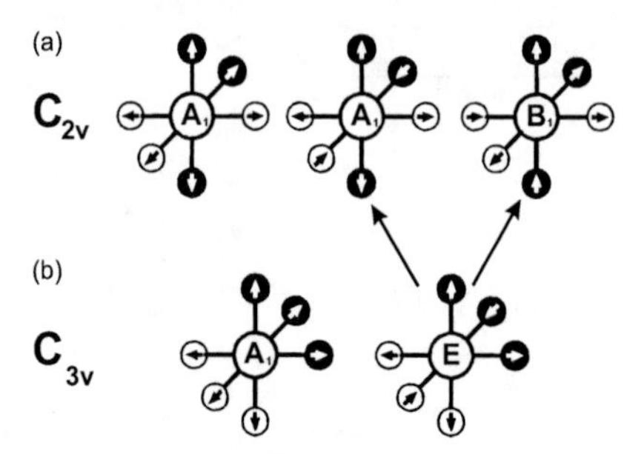

Figure 4.15: *Schematic picture of the central part of the meridional (a) and the facial (b) isomer of Alq*$_3$. *Hollow and filled circles around the central Al-atom represent oxygen and nitrogen atoms, respectively. The three stretching modes of the meridional molecule (*C_{2v}*-symmetry) and the two for the facial molecule (*C_{3v}*-symmetry, one is degenerated) are marked by arrows in the O and N atoms.*

α-Alq_3, which consists of the meridional isomer (C_{2v}), six stretching vibrations are expected, three involving Al-N and three involving Al-O modes (see Figure 4.15). As they are all dipole-allowed, they are observable by IR-spectroscopy. According to Kushto et al. the following assignments for α-Alq_3 are made: Al-N stretching: 396cm^{-1}, 405cm^{-1}, 418cm^{-1}, Al-O stretching: 522cm^{-1}, 542cm^{-1}, 549cm^{-1}. By contrast δ-Alq_3 shows a total of only four bands in this region (397cm^{-1}, 423cm^{-1}, 531cm^{-1}, 548cm^{-1}). As the AlO_3N_3 fragment of the facial isomer belongs to symmetry C_{3v}, six stretching vibrations are expected here too, but four of them belong to two degenerate vibrational states and therefore only four bands should be observed in IR-spectroscopy, as is the case for δ-Alq_3. The Al-N stretching is found at 397cm^{-1} and 423cm^{-1}, the Al-O stretching at 531cm^{-1} and 548cm^{-1}. The degeneracy of the first and last band is not present in the α-phase of Alq_3 (see Figure 4.14 and Figure 4.15), which consists of the meridional isomer. Two lines are observed at 400cm^{-1} and at 550cm^{-1} in α-Alq_3, in agreement with theoretical calculations of Kushto et al.[Kus00] This is strong evidence of the presence of the different isomers in δ- and α-Alq_3.

A further point to consider is the fact that vibrational studies of the polycrystalline phases of Alq_3 do not only involve the intramolecular characteristics but also the effect of crystallinity of the sample. We are interested in the intramolecular differences related to the symmetry of the molecule and thus wish to distinguish between these two effects. This would enable us to confirm that the splitting of the Al-N and Al-O stretching modes at $396cm^{-1}$ and $549cm^{-1}$, respectively, are due to the different molecular symmetry.

Therefore polycrystalline samples and very fine powders are compared in Figure 4.16. The upper trace shows the polycrystalline samples as prepared and the lower trace the same samples after being rubbed directly on the KBr pellets, which produced a fine powder. Lower crystallinity of the samples also reduces the background due to scattering, especially in the higher energy region above $600cm^{-1}$, resulting in symmetric peaks in the measured spectrum. These peaks are mainly caused by the hydroxyquinoline ligands and become sharper as a result of the lower crystallinity. Nevertheless, the energetic positions of the peaks are not affected by the crystallinity of the sample.

The experiment enables bands to be identified that are due to intermolecular interactions within the ordered crystalline environment, and these bands can be used to determine the degree of crystallinity of the samples measured. If these crystallinity bands are not observed any more, it can be assumed that the characteristic is dominated by the molecule itself and its symmetry.

From these measurements three particular bands related to crystallinity are observed in the spectra, namely at $1380cm^{-1}$, $750cm^{-1}$ and $531cm^{-1}$. The lines at $750cm^{-1}$ and $1380cm^{-1}$ lose their doublet character in the powdered samples. This can be attributed to the intermolecular loss of geometric correlation. The Al-O band at $531cm^{-1}$ of δ-Alq_3 shows a significantly higher intensity in the crystalline sample, but from theoretical considerations the intensity of the Al-O stretching mode for the facial isomer at $531cm^{-1}$ is expected to be very small [Kus00]. This is indeed observed in the spectrum of the fine powder in Figure 4.16a. It is well known that molecular ordering in a crystalline environment can change the relative intensities of related vibronic bands, for example due to dipole-dipole interaction [Per82, Rho61].

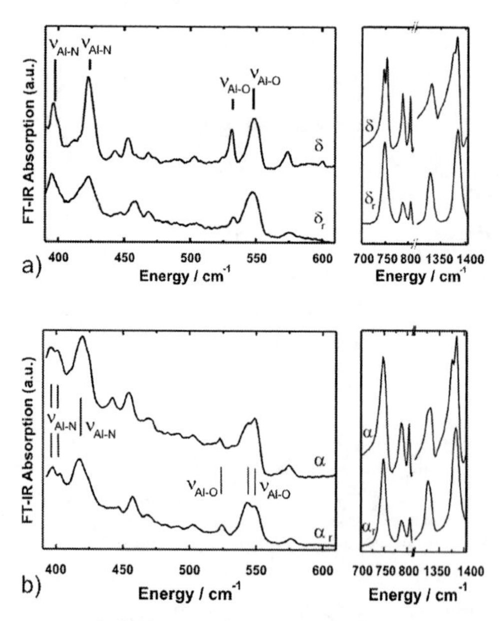

Figure 4.16: *Comparisons of polycrystalline Alq_3 as obtained from the sublimation tube and fine powders of the δ- and α-phases are shown in (a) and (b) respectively. The upper trace shows the polycrystalline samples as prepared and the lower trace the same samples after being rubbed mechanically directly on the KBr pellets, which produced a fine powder.*

As a result the three bands at 531cm^{-1}, 750cm^{-1} and 1380cm^{-1} serve as indicators for the influence of the packing in the crystal. It is clearly observed that in the powdered sample the doubling of the band centered at 750cm^{-1} in δ-Alq_3 disappears, confirming the assignment made in the first part of this section. The same holds for the double band observed for both phases at about 1380cm^{-1}. It also becomes a single mode after powdering. Consequently only the molecules without significant influence of the crystal structure are measured in the powdered samples, and this allows the criteria concerning the different isomerism of the molecules to be checked again: First of all, the Al-N and Al-O stretching vibrations found in the α- and δ- crystals in the region between

$390cm^{-1}$ and $600cm^{-1}$ remain, apart from slight shifts, at the same frequency after disintegration of the crystals. The degenerated bands of the facial isomer at $397cm^{-1}$ and $548cm^{-1}$ are still one peak for δ-Alq_3 (Figure 4.16(a)) and clearly split into two distinct peaks for the meridional α-Alq_3, respectively (see Figure 4.16(b)). Thus these differences between δ- and α-Alq_3 are not related to crystallinity or packing but are due to differences in the symmetry of the molecules. Second, although the vibrational peaks of both isomers become narrower after mechanical treatment, bands in α-Alq_3 remain broader than in δ-Alq_3, accounting for the inequivalence of the A, B and C ligands in the meridional Alq_3 molecule. Examples are the lines in the region of $640cm^{-1}$, $1333cm^{-1}$ and $1470cm^{-1}$ with FWHM of $17cm^{-1}$, $12cm^{-1}$, $13cm^{-1}$ and $20cm^{-1}$, $15cm^{-1}$, $19cm^{-1}$ for the δ- and the α-Alq_3, respectively. Therefore vibrational analysis is consistent with the interpretation that the meridional molecule constitutes the α-phase and the facial molecule is present in the δ-phase of Alq_3

In this section investigations of two different crystalline phases of Alq_3 (α-and δ-Alq_3) by using infrared spectroscopy are presented and it is demonstrated how the isomerism of the Alq_3 molecule is manifested in the vibrational properties. Comparison of the experimental results with theoretical calculations provides further evidence of the presence of the facial isomer in the blue luminescent δ-phase of Alq_3. Furthermore, the influence of intermolecular interactions due to the crystallinity of the sample is shown. It is to be expected that identification of the vibrational signatures of the facial isomer will allow further investigation of its properties and its role in OLEDs.

5 The Excited States of Alq_3

In the previous chapter structural investigations and properties of the molecule in the electronic ground state were discussed, giving evidence for the existence of the two different geometric isomers. However, not only the electronic ground state should be different for the two isomers, but also the excited states are expected to have different properties due to the different geometry of the molecule. Two types of photoexcited states are distinguished: the singlet state and the triplet state. In the singlet state the total spin quantum number of the unpaired electrons S=0, whereas in the triplet state the total spin quantum number is S=1.

The principal photophysical processes are summarized in the simplified schematic picture in Figure 5.1 [McG69]. Under electromagnetic radiation a molecule in the ground state (ψ_1) may be excited to a higher energetic state (ψ_2) if the energy hν of the incident photon corresponds to the difference between these two states ($h\nu=\Delta E=E(\psi_1)-E(\psi_2)$). This process is called absorption. The opposite process is light emission, when the molecule makes a transition from ψ_2 to ψ_1 under emission of a photon with the energy $h\nu=\Delta E$. Absorption and emission have a certain probability that is given by the oscillator strength f_{12}. Transitions with small f-values are often called forbidden transitions, whereas transitions with high f-values are called allowed transitions. In first approximation radiative transitions are allowed only between singlet states or between triplet states, and thus generally absorption implies a transition from the singlet ground state S_0 to a S_n state (n=1,2,...) or within the triplet states (e.g. $T_1 \rightarrow T_n$).

For emission, the emitting level of a given manifold is the lowest excited level of that manifold. This is called Kasha's rule, which is obeyed by most organic molecules. The state S_1 is the lowest excited singlet level and the emission from the $S_1 \rightarrow S_0$ transition is termed fluorescence, whereas the emission process from the lowest excited triplet level $T_1 \rightarrow S_0$ is termed phosphorescence.

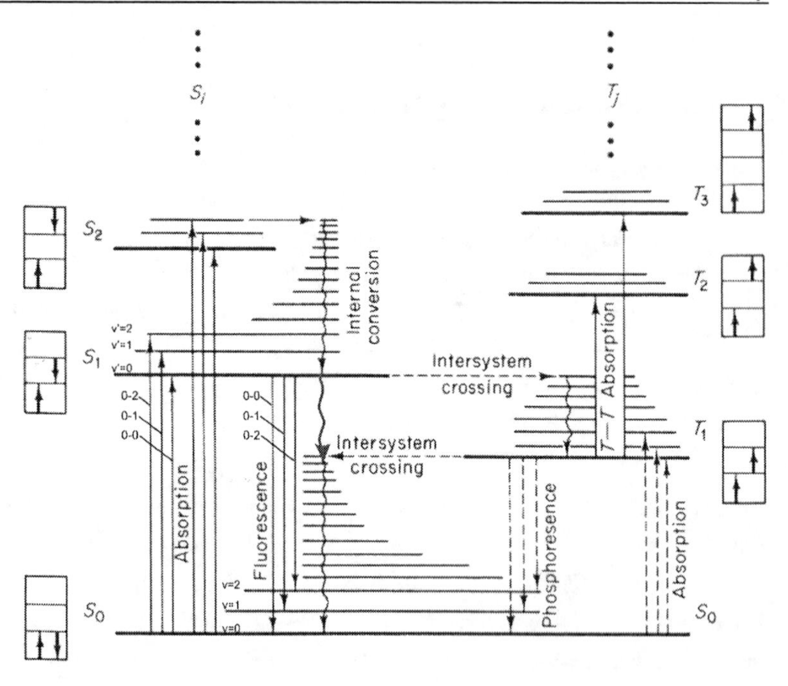

Figure 5.1: *Schematic level diagram for radiative and nonradiative decay processes in a typical organic molecule. Straight vertical lines symbolize radiative transitions. The dispositions of the electron spins are shown in the boxes (Diagram taken from S. McGlynn, T. Azumi and M. Kinoshita [McG69]).*

Kasha's rule also implies that higher excited states immediately relax to the S_1 or the T_1 state, respectively. This nonradiative process is called internal conversion. Another nonradiative process indicated in Figure 5.1 is intersystem crossing, describing the radiationless passage from an electronic state in the singlet manifold to an electronic state in the triplet manifold, or vice versa. This process is primarily responsible for the population of the triplet states in organic molecules under illumination, as the $S_0 \rightarrow T_n$ transition has a very low probability. The transition from the S_1 or the T_1 state to the S_0 ground state can also be radiationless by thermal conversion. The probability of an S_1 or T_1 state emitting a photon is given by the fluorescence or phosphorescence quantum yield, respectively.

In addition to these idealized "pure" electronic states, vibrational modes are coupled with these states as indicated by v and v' for the S_0 and S_1 states in Figure 5.1. In most cases absorption results in a transition from the lowest vibrational level (v=0) of S_0 to the vibrational modes (v'=0,1,2,...) of the excited S_1-state. Three of these transitions (0-0, 0-1, 0-2) are illustrated in Figure 5.1. The transitions leading to emission start from the vibrational ground state (v'=0) of the lowest excited state (S_1 or T_1) and end at vibronic levels of the ground state S_0. Three of these transitions from the S_1-state to the S_0-state and its coupled vibronic modes are also labeled in Figure 5.1 (0-0, 0-1, 0-2).

As in most cases the S_0-S_1 transition is an allowed transition, the lifetime of the S_1-state is very short. For Alq_3 it was measured to be about 12ns [Tan89, Mor97, Hum00, Kaw01]. On the other hand, the S_0-T_1 transition is a so-called forbidden transition and thus the lifetime of the T_1 state is expected to be several orders of magnitude larger. However, so far no experimental data on the triplet state of Alq_3 have been available, and thus this chapter presents the first measurements identifying and characterizing this excited state in Alq_3. The chapter is separated into two sections: The first reports investigations of the singlet excited state, whereas the second focuses on the triplet state of Alq_3 and its different phases.

5.1 Singlet Excited States

In order to obtain more insight into the nature of the singlet excited state, low temperature PL and PL excitation spectra were used. PL measurements give information about the transition from the S_1 state to the S_0 ground state, whereas PL excitation mainly sheds light on the $S_0 \rightarrow S_1$ transition. The results are discussed in terms of the isomerism of the molecule, and recently published theoretical calculations for the Alq_3 molecule are compared with the measured data.

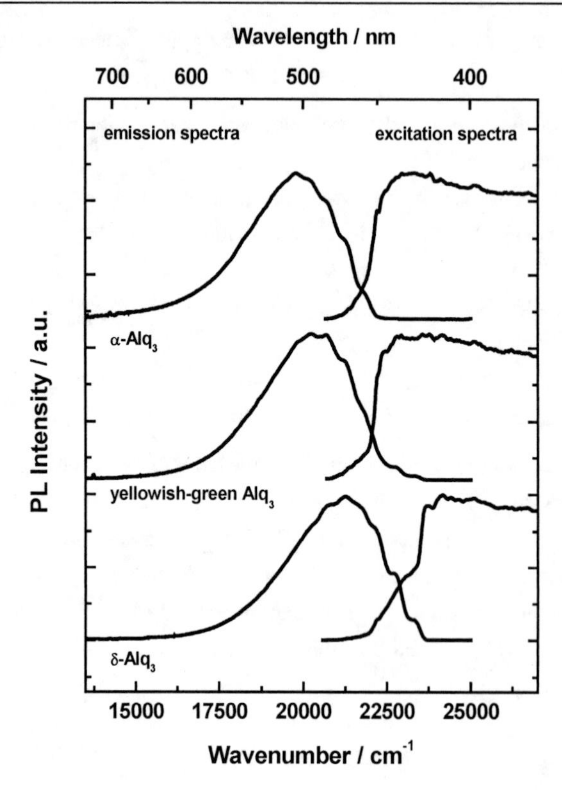

Figure 5.2: *Photoluminescence emission and excitation spectra of the different polycrystalline Alq_3 phases obtained from the sublimation tube. The emission spectra were taken at 1.3K by using excitation at 363.8nm. The excitation spectra were taken at 6K and the detection wavelength was 500nm in all cases.*

Results

PL excitation spectra and PL measurements in Figure 5.2 to Figure 5.4 give more insight into the nature of the fluorescence. Measured samples were taken from the sublimation tube as described in Chapter 4.1, and consequently the samples of fraction 1, fraction 2 and fraction 3 are named δ-Alq_3, yellowish-green Alq_3 and α-Alq_3, respectively.

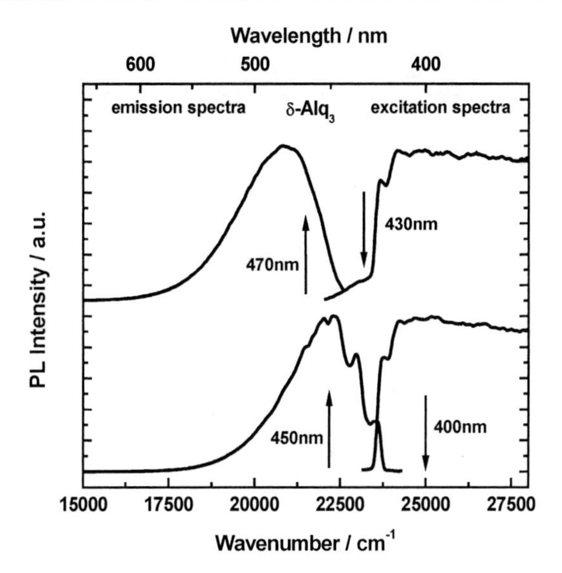

Figure 5.3: *Photoluminescence emission and excitation spectra of polycrystalline δ-Alq₃. Measurements were performed at 6K. The respective detection wavelength is labeled by up arrows and the excitation wavelength by down arrows. The PL spectrum excited at 430nm is significantly less intense than the spectrum excited at 400nm.*

The left of Figure 5.2 presents low temperature PL emission spectra for the different fractions excited at 363.8nm. There is a gradual shift of the PL maximum from 19800cm^{-1} (2.46eV) for α-Alq_3 to 21200cm^{-1} (2.63eV) for δ–Alq_3. The onset of these emissions also shifts from 2.74eV to 2.92eV. All PL spectra consist of a broad band, which is superimposed on the high energy side by a vibronic progression. Such well-resolved vibronic modes are not observed in amorphous films prepared by evaporation onto substrates at room temperature using any of these fractions as source material. The highest energetic vibronic band of α-Alq_3, which corresponds to the 0-0 transition, is at 21700cm^{-1} (2.69eV). The PL spectrum of δ-Alq_3 excited at 363.8nm also shows vibronic progression, but with a 0-0 transition at 23400cm^{-1} (2.9V) that is blue-shifted compared to α-Alq_3. Both phases show similar energetic differences between the observed vibronic bands of about 550cm^{-1}.

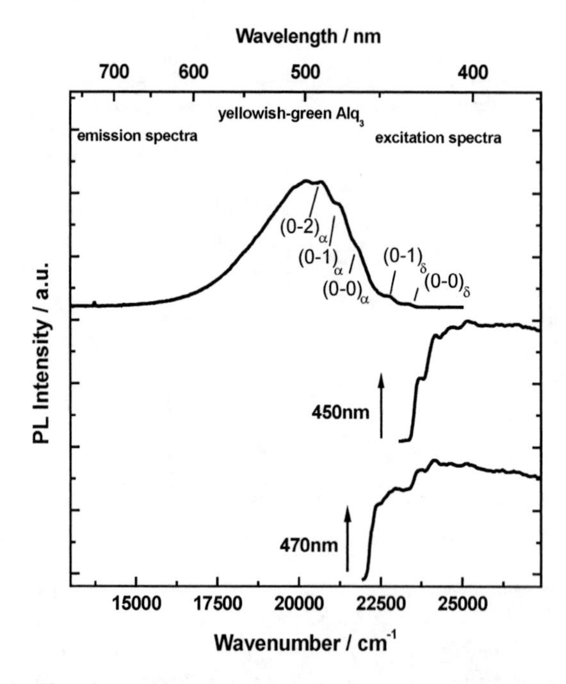

Figure 5.4: *Photoluminescence spectra of yellowish-green Alq_3 (fraction2). The emission spectrum was taken at 1.3K by using excitation at 363.8nm. The excitation measurements were performed at 6K. The respective detection wavelength is labeled by up arrows. The preliminary assigned vibronic modes belonging to the α- and δ- phase are labeled for clarity.*

The right of Figure 5.2 shows the PL excitation spectra of the three fractions detected at 500nm, located close to the PL maximum of α-Alq_3 and in the low-energy tail of the PL spectrum of δ-Alq_3. The excitation spectra of α-Alq_3 and yellowish-green Alq_3 are very similar with the same energetic position of the excitation edge at 455nm (22050cm^{-1}), but for δ-Alq_3 an additional blue-shifted excitation edge is observed.

In order to demonstrate the behavior of the δ-phase in detail, PL and PL excitation spectra of this phase are shown in Figure 5.3. The PL excitation spectra of δ-Alq_3 detected at 450nm and 470nm show a sharp excitation edge at

23400cm^{-1} (2.92eV) with well-resolved vibronic modes. The PL spectrum excited at 400nm also shows vibronic progression and a small Stokes shift of only 200cm^{-1} (25meV) is observed. The vibronic progression in the PL spectrum is seen only for excitation energies above the sharp excitation edge. Upon excitation below this edge, for example at 430nm, as shown in Figure 5.3, the vibronic progression in the PL disappears and the resulting spectrum is characterized by a broad band which is still blue-shifted relative to the PL of the other fractions. Further decrease of the excitation energy leads to an increasing shift of the PL maximum to lower energies. The intensity of these PL spectra is significantly less compared to excitations above the excitation edge, for example at 400nm. Such behavior can be described by involving carrier thermalization in lower-energy tail states similar to the case of amorphous films [Hum00, Tzo01]. The existence of states below the excitation edge was confirmed by excitation spectra detected at lower energies and also becomes obvious from the PL excitation spectrum of δ-Alq_3 detected at 500nm, as shown in Figure 5.2. Further evidence for the existence of tail and subgap states was found by absorption measurements on polycrystalline samples using photothermal deflection spectroscopy [Tzo01] and photoinduced absorption measurements, the discussion of which is beyond the scope of this work. The broad feature below the excitation edge becomes more pronounced at lower detection energy, which confirms the heterogeneous character of the photoluminescence.

In Chapter 4.1 yellowish-green Alq_3 was found to be a mixture of the α- and δ-phase. This was also investigated using PL and PL excitation spectra. The PL spectrum of yellowish-green Alq_3 given in Figure 5.2 and Figure 5.4 shows vibronic bands with energetic positions typical of α-Alq_3. However, there are two weak peaks at the high energy side of the spectrum, which are similar to the δ-phase spectrum. This finding is confirmed by the excitation spectra in Figure 5.4. Detection at the high energy side (450nm) of the PL band results in an excitation spectrum with an edge like that for δ-Alq_3, whereas detection at lower energy (470nm) gives an excitation edge red-shifted by 1400cm^{-1} (0.17eV) like for α-Alq_3. This allows us to conclude that yellowish-green Alq_3, which is very often used for device fabrication, is constituted mainly of α-Alq_3 with inclusions of δ-Alq_3, which is consistent with the results in Chapter 4.1. Another important finding is that α-Alq_3 and δ-Alq_3 are characterized by energetically displaced 0-0 vibronic bands in addition to the displacement of the overall PL maximum.

Discussion

The experimental results from PL and PL excitation spectra in this section give further evidence that the δ-phase consists of molecules in the facial isomer state. The most significant difference in the measured spectra is the blue shift of the δ-phase compared to the α-phase. Recently Amati et al. [Ama02a, Ama03] and Curioni et al. [Cur98] published results of TD-DFT simulations and ab initio calculations, respectively, for the two different isomers. In these studies the energy gap for both isomers was calculated to be between 0.13eV and 0.3eV larger in the facial isomer. This value is indeed comparable with the blue shift of about 0.2eV observed for the maximum of the PL and the 0-0 transition of δ-Alq_3 relative to α-Alq_3. A more detailed comparison of our experimental data with theoretical calculations of Amati and Lelj is discussed below.

The origin of the observed vibronic progression can be found in the coupling of the electronic transitions to vibrational modes with energies around $550cm^{-1}$, as was also discussed by Brinkmann et al. for the α-phase of Alq_3 [Bri00]. From vibrational analysis in Chapter 4.4 it can be seen in Table 4.3 that there are several modes around this energy. Therefore a unique assignment is not possible, although this energetic region is dominated by Al-O stretching modes.

Amati and Lelj published theoretical calculations for the meridional and the facial isomer of Alq_3 using time-dependent density functional theory (TD-DFT) [Ama02a, Ama03]. They computed the excitation energies and oscillator strengths of the excited singlet states for the meridional isomer and for the facial isomer, respectively. The distribution of the orbitals on the ligands and the energies of molecular orbitals close to the HOMO and LUMO levels as shown in Figure 2.2 were obtained. These results are similar to ab initio calculations published by Curioni et al. [Cur98]. From these data the energetic distances between ground states and excited states were obtained. These energetic distances give the wavelengths where optical absorption is expected for the two geometric isomers. Table 5.1 summarizes the results for the first three energetic transitions calculated by Amati et al. [Ama02a, Ama03].

Table 5.1: *Computed singlet-singlet excitation energies (in eV and nm) and oscillator strengths for the first three excitations of the facial and meridional isomer, respectively. For details of these calculations made by Amati and Lelj see references [Ama02a] and [Ama03]. For facial Alq$_3$ the oscillator strength of the first excited state is very strong, whereas in meridional Alq$_3$ it is very weak and the second excited state has the dominant oscillator strength.*

Facial Alq$_3$				**Meridional Alq$_3$**			
Excited state	**Transitions**	**Energy / eV (Wavelength / nm)**	**Oscillator strength**	**Excited state**	**Transitions**	**Energy / eV (Wavelength / nm)**	**Oscillator strength**
1E	II-e→III-e II-e→III-a II-a→III-e	2.89 (429)	**0.0588**	1	II-C→III-B II-A→III-B	2.77 (447)	**0,0052**
1A	II-e→III-e	2.93 (423)	0.0061	2	II-B→III-B II-C→III-B II-A→III-C II-A→III-A	2.90 (427)	0.0671
2E	II-e→III-e II-e→III-a II-a→III-e	3.00 (414)	0.0083	3	II-C→III-B II-A→III-B II-A→III-C II-A→III-A	2.94 (422)	0.0021

Further important results of their calculations are the obtained oscillator strengths for the lowest energy transitions, also given in Table 5.1. The oscillator strength is proportional to the probability of the corresponding energy transition. Here we find an important difference between the two isomers in the oscillator strength of the lowest energy transition. For the facial isomer, the lowest energy excitation is predicted to be the most probable (greatest oscillator strength) and from this a comparatively steep absorption edge is expected. On the other hand, for the meridional isomer, the second excitation appears much more intense than the first. Approximately the same oscillator strengths were also calculated by Martin et al. for the meridional isomer (0.006 and 0.068 for the first and second transition, respectively) [Mar00].

The theoretical calculations for the molecular orbits of the meridional and facial Alq$_3$ molecule, which determine differences in the excitation energies of their first excited states as well as different relative oscillator strengths for the first and second excitation, give very interesting conclusions for the measured results. They allow us to discuss whether the experimental data is in agreement with the presence of the facial isomer in the δ-phase of Alq$_3$.

Vibrational relaxation is predicted to be similar for both isomers, as the molecular orbitals involved in the first transition (set II and set III in Figure 2.2) are due to similar fragment orbitals in the two isomers. Furthermore, the fragment orbitals are localized on the ligands and thus the vibrational relaxation of the first excited state is expected to be approximately the same (about $500cm^{-1}$) for both isomers, as observed in the measured spectra [Ama02a].

In our case PL excitation spectra instead of absorption spectra were used to compare the two phases, as the defined thickness-controlled preparation of thin homogeneous films of δ-Alq_3, which are required for absorption measurements, needs further improvement. The first excited state can be measured by using PL excitation spectra of sufficiently thick solid samples. This results from the almost complete absorption of the incident light due to the great sample thickness. With this method the weak absorption of the first excitation of the meridional isomer in α-Alq_3 can be measured although it is much less probable than the second transition, and it allows us to assign the excitation absorption edge to the lowest electronic transition. For δ-Alq_3 a sharp excitation edge is observed at 426nm in Figure 5.3. This is in very good correspondence with the computed excitation at 429nm. Furthermore, the absorption edge of δ-Alq_3 is about $1400cm^{-1}$ (0.17eV) blue-shifted relative to the α-phase. The computed difference in energy between the lowest excited state of each isomer is about $1060cm^{-1}$ (0.13eV), which is in reasonable agreement with the experimental result, and therefore the energetic position as well as the measured blue-shift suggest that the δ-phase consists of the facial isomer. Thus it can be concluded that all experimental data, namely the energetic position of the $S_0 \rightarrow S_1$ transition of δ-Alq_3 and the blue-shift of its excitation edge, strongly support the identification of the facial isomer in δ-Alq_3.

5.2 The Triplet State in Alq_3

In this section the first experimental observation of the triplet state in Alq_3 is reported and the zero field splitting parameters E and D introduced in Chapter 3 are measured. After that further characteristic properties of the triplet state are determined by measuring their lifetime and estimating their energy as well as the population of the triplet state due to intersystem crossing. This is performed for the different phases as well as for evaporated amorphous films. The discussion further addresses the question whether the experimental results are consistent with the facial isomer in δ-Alq_3.

Results

The first objective was to gain experimental evidence for the existence of triplets in Alq_3 since no experimental data on the triplet state of Alq_3 have been available so far. One method to identify triplet states is optically detected magnetic resonance (ODMR) described in Chapter 3. Figure 5.5 shows ODMR signals at zero field of α-Alq_3, yellowish-green Alq_3, δ-Alq_3 and an amorphous film detected at the respective wavelength of their PL maximum. One clearly observes signals at about 700MHz, 1500MHz and 2200MHz, which confirms the presence of the triplet state in all phases. The energetic position of the signals is very similar for all phases and is only slightly shifted for the amorphous film; thus it can be concluded that it is independent of the morphology. From these measurements the zero field splitting parameters E and D can be evaluated and are almost identical for all phases: |E| is in the range of $0.0114cm^{-1}$ and |D| is about $0.0630cm^{1}$ (see Table 5.2). The concentration of triplet states in δ-Alq_3 seems to be low compared to the other samples because the ODMR signal is very weak, as can be seen from the strong noise in Figure 5.5. The ODMR measurements clearly identified the triplet state in Alq_3, and we found that their characteristic values are similar for all phases.

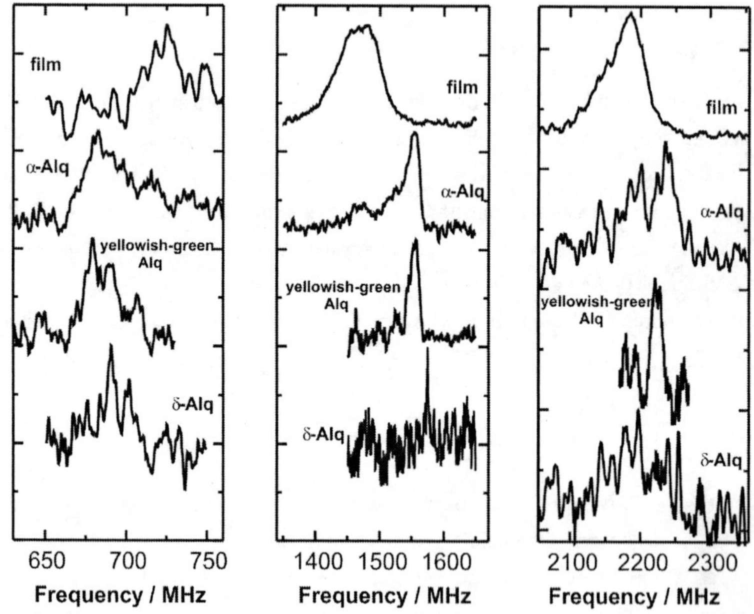

Figure 5.5: *ODMR signals of the three fractions obtained from the sublimation tube (α-Alq_3, yellowish-green Alq_3 and δ-Alq_3), and of an evaporated film (thickness: 1µm) at 1.3K. The signal was detected at the maximum of their PL intensity in each case.*

To learn more about the properties of the triplet state, measurements of the transient PL were used. From these it is possible to obtain information about the lifetime and the population of the triplet state due to intersystem crossing. The principle of these measurements is shown in Figure 5.6. The sample is excited by a rectangular laser pulse and instantaneously with the turning-on of the excitation light fluorescence is observed, which subsequently decreases to an equilibrium value with a decay time of about 10ms. The decay is related to the intersystem crossing as will be discussed in detail below. In order to be sure of having reached this dynamic equilibrium value at the time the laser is turned off, the excitation pulse was chosen to be about 30-40ms. After the laser is turned off, the intensity of spontaneous PL from the singlet states decreases very fast due to their short lifetime of about 12ns [Tan89, Mor97, Hum00, Kaw01]; thus after just 1µs no spontaneous fluorescence is present any more. However, even after a

Table 5.2: *Values obtained from the ODMR measurements for the different Alq_3-phases.*

	ODMR-signal in MHz	**ODMR-signal in MHz**	**ODMR-signal in MHz**	**|E| in (MHz) and cm^{-1}**	**|D| in (MHz) and cm^{-1}**
δ-Alq_3	680±20	1560±20	2230±20	(340±10) 0.0114±0.0003	(1895±40) 0.0632±0.0013
yellowish-green Alq_3	680±10	1550±10	2230±10	(340±5) 0.0114±0.0002	(1890±20) 0.0630±0.0007
α-Alq_3	680±10	1555±10	2235±10	(340±5) 0.0114±0.0002	(1895±20) 0.0632±0.0007
film	720±10	1465±10	2185±10	(360±5) 0.0120±0.0002	(1825±20) 0.0609±0.0007

delay time Δt of 2ms a weak luminescence is still observed, as shown in Figure 5.6. It is about three orders of magnitude less intense than the spontaneous PL and shows a slow decay rate.

This process is known as delayed fluorescence. Whereas the singlets have a short lifetime (only in the range about 10ns) and thus relax instantaneously, the triplet states live much longer and their lifetime in organic materials is often in the range of several ms. Delayed fluorescence occurs due to collision of two triplet excitations and is therefore a bimolecular process, which has a probability proportional to the square of the density of the triplet states $[T_1]$. The collision process is also often referred to as triplet exciton fusion. If the energy of the lowest excited singlet state is less than the sum of the energies of the colliding triplet excitons, the fusion reaction may yield triplet and singlet states. The possible reactions can be written as

$$\begin{aligned} & T_1 + T_1 \rightarrow S_0 + T_n \rightarrow S_0 + T_1 \\ & \text{or} \\ & T_1 + T_1 \rightarrow S_0 + S_n \rightarrow S_0 + S_1 \xrightarrow{h\nu} S_0 + S_0 \end{aligned} \quad (5.1)$$

where T_n and S_n represent a particular vibrational state in the n-th triplet and n-th singlet electronic manifolds, respectively. The second equation describes the

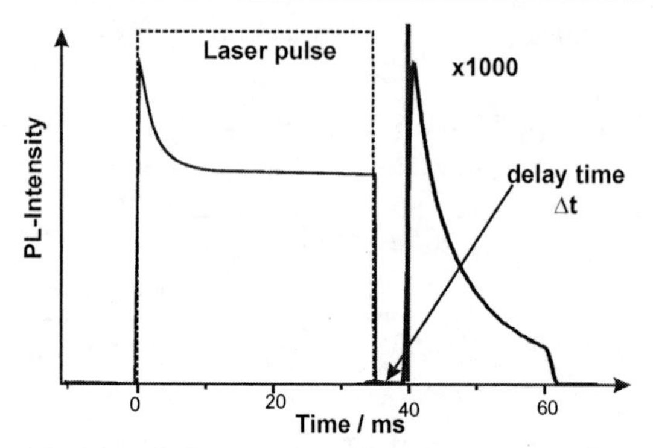

Figure 5.6: *Principle of the transient PL measurements shown for α-Alq_3 measured at 20K. By applying an intense rectangular laser pulse, PL from the sample is observed instantaneously. During the first 10-15ms the PL-signal decreases due to intersystem crossing to a constant equilibrium value. After the laser is turned off, the spontaneous PL decreases within less than 1μs. The observed fluorescence after the laser has been turned off is called delayed fluorescence, with an intensity of about three orders of magnitude less than the spontaneous PL. In our experiments this delayed fluorescence was measured after a delay time Δt of several milliseconds. The cut-off at about 60ms in the schematic picture is due to the chopper system.*

process leading to delayed fluorescence. The singlet exciton S_n subsequently relaxes to the lowest excited singlet exciton state and decays as would be the case for a directly excited singlet exciton state. The distinguishing feature of the fluorescence from the S_1-state generated by the fusion process is that its apparent lifetime is determined by the triplet excitons and hence is much longer than the spontaneous fluorescence lifetime.

As the delayed fluorescence originates from the $S_1 \rightarrow S_0$ transition, its spectrum is expected to be the same as observed in usual PL measurements. Figure 5.7 shows delayed fluorescence spectra of α- and δ-Alq_3 taken between 2 and 5ms after the end of the laser pulse. The samples were excited with a power of 10mW at 363.8nm and with a focused laser beam in order to have maximized excitation density. Both phases show clearly resolved spectra, which are identical to those obtained by usual PL measurements shown in Figure 5.2, but by a factor of 500 to 1000 less intense. In addition the signal intensity of the δ-phase was less than

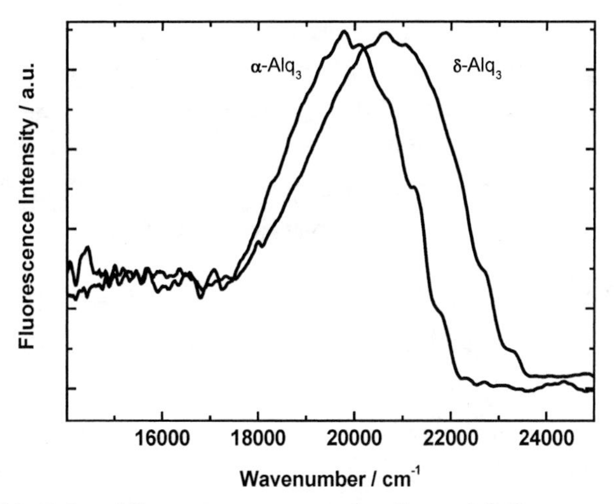

Figure 5.7: *Delayed fluorescence spectra of α-Alq_3 and δ-Alq_3 measured after a delay time Δt of 2-4ms at a temperature of 6K. The samples were excited at 363.8nm using a laser beam with a power of 10mW, which was focused in order to have a maximized excitation density. Both spectra are similar to the PL spectra in Figure 5.2 but with a weak additional signal on the low energy side.*

that of the α-phase. Therefore the spectrum is due to the $S_1 \rightarrow S_0$ transition as expected. On the low energy side of the spectrum a pronounced tail or additional band is observed. Its relative intensity increases compared to the PL maximum by increasing the delay time Δt; thus it has a longer lifetime than the delayed fluorescence and must be of somehow different origin.

To obtain further information about the triplet states and the origin of the additional low energy band, more detailed investigations of the delayed luminescence were made. For these measurements only a He-Cd laser was available, from which the intense line at 442nm was used. Figure 5.8 shows the delayed luminescence spectra of α-Alq_3, yellowish-green Alq_3, δ-Alq_3 and an evaporated amorphous film at 10K with a delay time Δt of 4ms; the spectrum was integrated over about 20ms. All spectra show two distinct bands, one at about 500nm, similar to spontaneous PL, and an additional band at about 700nm. For α-Alq_3, yellowish-green Alq_3 and the evaporated film the position of the bands is approximately the same, whereas for δ-Alq_3 the band around 700nm is blue-shifted. The relative intensity of the two bands is different for the different phases and temperature-dependent, as will be discussed in detail in the diploma thesis of Christoph Gärditz [Gär03] and in Ref.[Cöl04b]. As the wavelength of 442nm is located in the tail of the absorption of α-Alq_3 and in particular below the absorption edge of δ-Alq_3 (see Figure 5.3) the obtained density of excited states is significantly lower than by excitation at 363.8nm. This explains the differences of the delayed fluorescence spectra in Figure 5.8 compared to Figure 5.7, especially for the δ-Alq_3 sample. Under these experimental conditions only a very weak band at 500nm is observed for δ-Alq_3 due to the low density of excited states, but the band at 700nm is still clearly resolved.

For all phases vibronic progressions on the high energy side of the band at 700nm are observed. We note that this and the following results are also valid for phosphorescence measured from Alq_3-OLEDs [Cöl04a]. Subtracting the usual PL spectrum from the spectra in Figure 5.8 gives nearly identical bands at 700nm with vibronic progressions at approximately the same positions for the film, α-Alq_3 and yellowish-green Alq_3. The vibronic progressions are located at about 586nm, 606nm, 627nm, 645nm and 668nm (17065cm^{-1}, 16502cm^{-1}, 15950cm^{-1}, 15504cm^{-1} and 14970cm^{-1}) and for δ-Alq_3 at about 574nm, 594nm, 613nm and 635nm (17422cm^{-1}, 16835cm^{-1}, 16313cm^{-1} and 15748cm^{-1}) and thus have an average distance of about 550cm^{-1}, similar to the vibronic progression observed for the PL in the previous section. Therefore the vibrational modes of the new band at about 700nm seem to be due to the vibrational modes of the Alq_3 molecule in its electronic ground state.

The transient luminescence of both bands was also investigated. By measuring the decay of the delayed fluorescence intensity I_{DF}, the lifetime of the triplet states can be determined [Pop82, Kao81]. As the lifetime of the singlet excitons is very short compared to the triplet excitons, the spontaneous fluorescence decays very fast after the end of the optical excitation and within less than 1μs there is a dynamic equilibrium between the generation of S_1-states due to triplet fusion and the decay of S_1-states. This can be written as

$$k_s[S_1] = \tfrac{1}{2} f \gamma_{tot} [T_1]^2 .$$

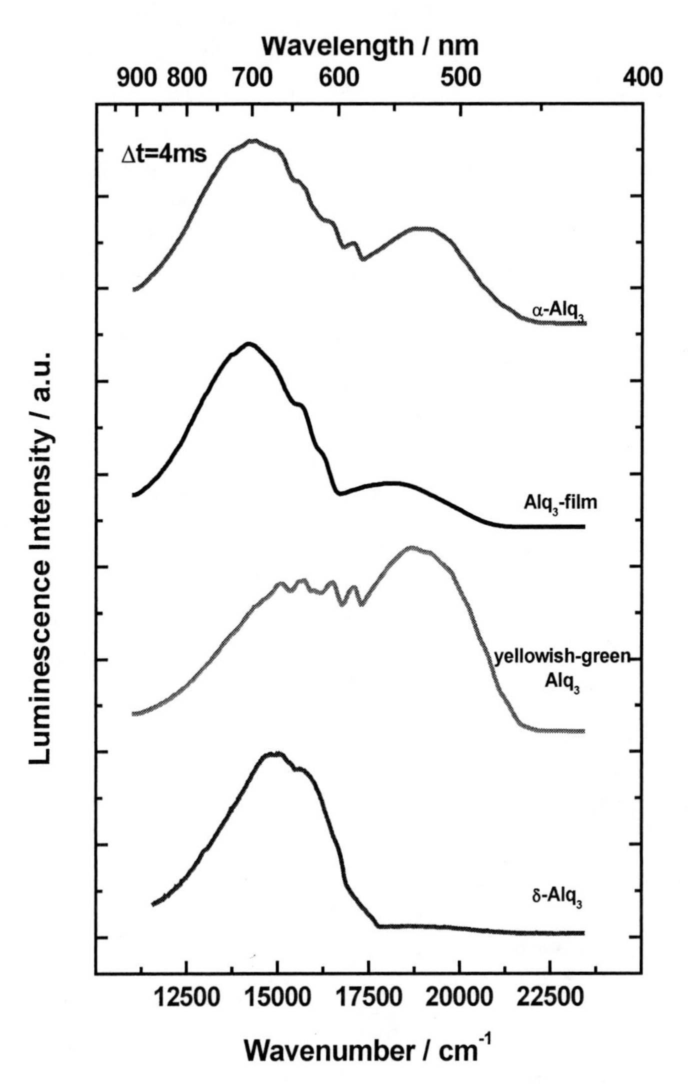

Figure 5.8: *Delayed fluorescence spectra of Alq_3 taken after a delay time Δt of 4ms after the end of the laser pulse and integrated over about 20ms at 10-20K. Excitation wavelength was 442nm. All samples show two distinct bands: the typical PL spectrum and a new additional band at about 700nm (1.77eV). This new band is blue-shifted for δ-Alq_3.*

k_s is the rate constant for the decay of the singlet states and $[S_1]$ is the density of the S_1-states. f is the fraction of triplet-triplet annihilations that leads to a singlet exciton, γ_{tot} is the total bimolecular annihilation (fusion) rate constant, and the factor of ½ occurs because the disappearance of two triplets results in only one singlet. As triplet-triplet annihilation is a bimolecular process, it is proportional to the square of the triplet concentration. The fluorescence intensity is proportional to the concentration of the singlet states, with k_r giving the probability of radiative decay, and thus the intensity of the delayed fluorescence I_{DF} is

$$I_{DF} = k_r[S_1] = \Phi_F \tfrac{1}{2} f \gamma_{tot} [T_1]^2 \tag{5.2}$$

where the constant Φ_F is the fluorescence quantum efficiency. Starting with an initial value $[T_1]_0$, the triplet concentration after the excitation light has been turned off (t=0), the time evolution of the concentration of the triplet state $[T_1]$ is given by

$$\frac{d[T_1]}{dt} = -k_T[T_1] - \gamma_{tot}[T_1]^2, \tag{5.3}$$

where k_T is correlated with the triplet lifetime by $k_T=1/\tau_0$. Solving this equation gives the time dependence of $[T_1]$ and thus the time dependence of I_{DF}.

At very high triplet concentrations it is possible that the second term in equation (5.3) is the dominant process, which reduces the population of the triplet states, and equation (5.3) becomes

$$\frac{d[T_1]}{dt} = -\gamma_{tot}[T_1]^2.$$

The solution of this equation is

$$[T_1] = \left(\frac{1}{[T_1]_0} + \gamma_{tot} t \right)^{-1},$$

and thus with equation (5.2):

$$\frac{1}{\sqrt{I_{DF}(t)}} = \frac{1}{\sqrt{I_{DF}(t=0)}} + \frac{1}{\sqrt{\Phi_F \frac{1}{2} \frac{f}{\gamma_{tot}}}} t$$

$$\Rightarrow \frac{1}{\sqrt{I_{DF}(t)}} \sim t$$

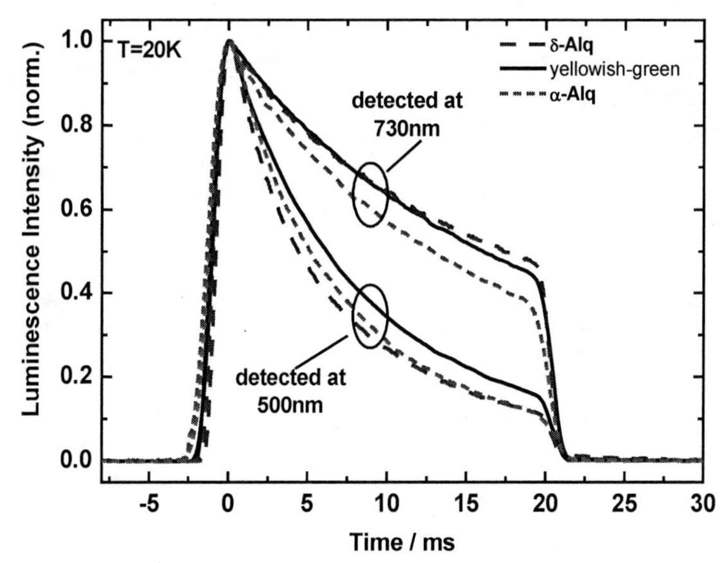

Figure 5.9: *Transient intensity of the delayed luminescence shown in Figure 5.8 detected at 500nm and of the band with the maximum at 700nm detected at 730nm. The delay time Δt was 4ms in all cases. The steep edge at 20ms is due to the experimental setup. The temperature was 20K.*

This time dependence was observed in anthracene crystals immediately after intense UV-excitation [Hal63]. However, after some time (some milliseconds) or by using less intense excitation light, the second term in equation (5.3) can be neglected due to the small triplet concentration ($k_T[T_1] >> \gamma_{tot}[T_1]^2$):

$$\frac{d[T_1]}{dt} = -k_T[T_1]$$

and thus

$$[T_1] = [T_1]_0 e^{-k_T t}$$

and with equation (5.2)

$$I_{DF}(t) \sim e^{-2k_T t}. \quad (5.4)$$

Therefore I_{DF} is a monoexponential decay. The decay time of the delayed fluorescence intensity is half of the correlated triplet lifetime τ_0 ($\tau_0 = 1/k_T$), and

Table 5.3: *Lifetimes obtained from the transient measurements of the delayed fluorescence intensity of the different polycrystalline Alq_3 phases and an amorphous film at a temperature of about 20K. τ_{DF} is the time constant for the exponential decay of the delayed fluorescence intensity and from this the triplet lifetime τ_0 was obtained. τ_{700} is the exponential decay of the emission measured at 730nm.*

	τ_{DF}	τ_0 (=$2\tau_{DF}$)	τ_{700}	τ_{700}/τ_{DF}
α-Alq_3	6.6±0.5	13.2±1	13.6±0.5	2.05
yellowish-green	7.8±0.5	15.6±1	16.2±0.5	2.08
δ-Alq_3	6.2±0.5	12.4±1	13.2±0.5	2.13
film	4.33±0.5	8.66±1	9.3±0.5	2.15

thus it is possible to determine the lifetime of the triplet state with transient measurements of the delayed fluorescence intensity.

Figure 5.9 shows the intensity decay of the delayed luminescence of polycrystalline samples detected at 500nm and of the additional band detected at 730nm and measured at a temperature of 20K. All samples show a very good monoexponential decay. Therefore the measurements were clearly in the regime where equation (5.4) has to be used. All three samples in Figure 5.9 show a similar monoexponential decay for the delayed fluorescence detected at 500nm. The measured apparent lifetimes are 6.6ms±0.5, 7.8±0.5ms and 6.2±0.5ms, resulting in triplet lifetimes of 13.2±1ms, 15.6±1ms and 12.4±1ms for α-Alq_3, yellowish-green Alq_3 and δ-Alq_3, respectively. The values are summarized in Table 5.3. Although in the film the triplet lifetime is about 60%-70% of that in the polycrystalline samples, all values are in the same range and thus the morphology of the samples seems to have only little influence on the lifetime of the triplet states.

The decay of the new band at about 700nm is also shown in Figure 5.9. All polycrystalline phases show a similar monoexponential decay, which is significantly slower than that detected at 500nm. Measured monoexponential

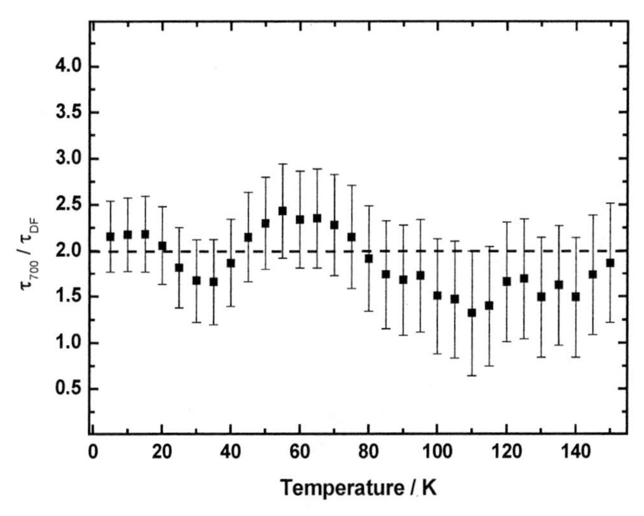

Figure 5.10: *Temperature dependence of the factor (τ_{700}/τ_{DF}) between the decay times measured for the delayed fluorescence at 500nm (τ_{DF}) and for the band at 700nm (τ_{700}) The dashed line at the factor(τ_{700}/τ_{DF})=2 serves as a guide to the eye.*

decay times are also given in Table 5.3. Within the accuracy of the measurement these values are about a factor of 2 higher than the values for the band at 500nm and thus almost identical to the triplet lifetimes obtained. The temperature dependence of the decay of both bands was investigated for all phases of Alq_3 [Gär03, Cöl04b], and as a result it became clear that both bands are directly correlated. As shown in Figure 5.10, the lifetime of the band located at about 700nm and the apparent lifetime of the delayed fluorescence always differ by a factor of 2 within the accuracy of the measurements, and thus the lifetime of the new band is the lifetime of the triplet state. From this it is clear that this band is directly linked with the triplet state of Alq_3.

The triplet state is populated due to intersystem crossing, as schematically shown in Figure 5.1 and Figure 5.11. Due to photoexcitation by the absorption of incident laser light, mainly the singlet states S_n are excited ($S_0 \rightarrow S_n$) and relax to the lowest excited singlet state S_1 (process a). The triplet state is populated by intersystem crossing with the rate constant d. Under constant photoexcitation and

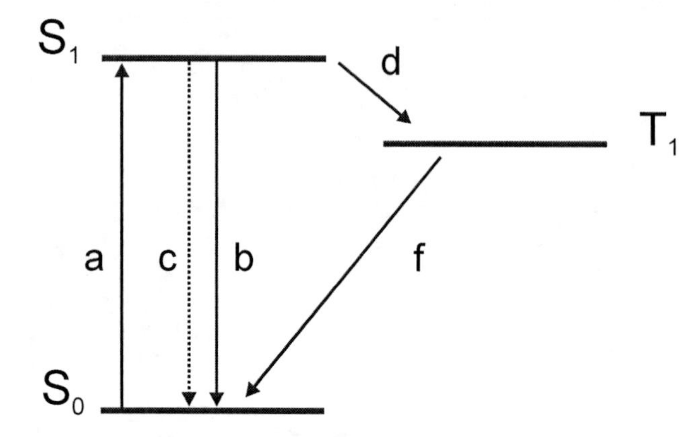

Figure 5.11: *Schematic diagram of the electron levels and the transitions between the levels in an organic molecule. S_0, and S_1 are the non-excited ground state and the first excited singlet level; T_1 is the lowest triplet level. The coefficient a is proportional to the intensity of the exciting light and the probability of excitation of the molecule. b, c, d and f are the corresponding rate constants. The transition d is the population of the triplet state due to intersystem crossing.*

for long times ($t\rightarrow\infty$) there is a dynamic equilibrium of the $S_1\rightarrow T_1$ and $T_1\rightarrow S_0$ transitions, resulting in a constant concentration of the triplet states $[T_1]^\infty$. The molecules which are in the long-lived triplet state are not able to emit fluorescent light and, at high excitation density, this leads to a decrease in fluorescence intensity. Therefore the process of intersystem crossing can be investigated by transient PL measurements and as a result the ratio of molecules in the triplet state can be estimated. The time dependence of the population process and the concentration of the triplet states $[T_1]^\infty$ is obtained from the rate equations (see Figure 5.11):

$$\frac{d[S_0]}{dt} = -a[S_0] + b[S_1] + c[S_1] + f[T_1] \tag{5.5}$$

$$\frac{d[S_1]}{dt} = a[S_0] - b[S_1] - c[S_1] - d[S_1] \tag{5.6}$$

$$\frac{d[T_1]}{dt} = d[S_1] - f[T_1] \tag{5.7}$$

The definition of the rate constants *a*, *b*, *c*, *d* and *f* is shown in Figure 5.11. In the literature *f* is often denoted k_T. These equations were solved by Sveshnikov, and Smirnov et al. [Sve48, Smi66]. The constants *b* and *c* can be replaced by $b'=b+c$. Further, if we bear in mind both that the lifetime of the triplet state is much longer than the lifetime of the singlet state and that the rate of intersystem crossing is much higher than the rate of triplet decay ($b'>>f$, $d>>f$), the solutions are

$$[S_1]=\frac{af[S_0]^0}{(b'+d)f+da}+\frac{Aa^2[S_0]^0}{(b'+d)f+da}e^{-\frac{1}{\tau_1}t}-\frac{a[S_0]^0}{b'+d+Ba}e^{-\frac{1}{\tau_2}t} \tag{5.8}$$

$$[T_1]=\frac{ad[S_0]^0}{(b'+d)f+da}\left(1-e^{-\frac{1}{\tau_1}t}\right) \tag{5.9}$$

with

$$f+\frac{d}{b'+d}a=f+Aa=\frac{1}{\tau_1} \tag{5.10}$$

and

$$b'+d+\frac{b'}{b'+d}a=b'+d+Ba=\frac{1}{\tau_2}. \tag{5.11}$$

From these equations it is evident that τ_1 is the characteristic time for the accumulation of molecules in the triplet state. For $t\to\infty$ (stationary regime) the concentration of molecules in the triplet state is given by

$$[T_1]^\infty=\frac{ad[S_0]^0}{(b'+d)f+da}=Aa[S_0]^0\tau_1 \tag{5.12}$$

and finally with equation (5.10) one can express $[T_1]^\infty$ by the characteristic accumulation time τ_1 and the lifetime of the molecules in the triplet state $\tau_0=1/f$.

$$[T_1]^\infty=[S_0]^0\left(1-\frac{\tau_1}{\tau_0}\right) \tag{5.13}$$

As a result it is possible to estimate the ratio of the molecules excited in the triplet state $[T_1]^\infty$ from the lifetime τ_0, determined from the delayed fluorescence

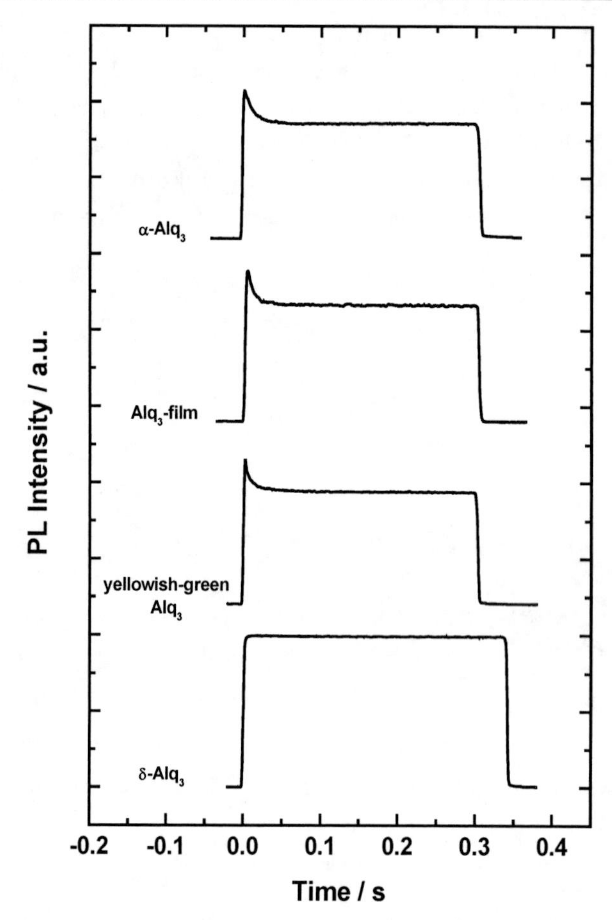

Figure 5.12: *Time dependence of the PL-intensity during an optical excitation pulse for polycrystalline Alq_3 phases and an evaporated film. The measurements were performed at 1.3K by using excitation at 363.8nm. The signal was detected at 2.64eV (470nm) for δ-Alq_3 and at 2.48eV (500nm) for all other samples.*

measurements, and the characteristic accumulation time τ_1, which can be measured using transient PL studies.

These measurements for the Alq_3 phases as well as for an evaporated amorphous film are shown in Figure 5.12. Instantaneously with the turning-on of the excitation light the fluorescence is observed, which subsequently decreases with a decay time τ_1 to an equilibrium value. α-Alq_3, yellowish-green Alq_3 and the evaporated amorphous film behave in a similar fashion. Their decay time τ_1 is 11ms, 11ms, and 7ms, and the triplet lifetime τ_0 at that temperature (1.3K) was measured as 15ms, 14ms, and 9ms respectively. Therefore in these samples about 20% to 30% of the molecules are in the triplet state. This similarity between these samples of Alq_3 containing the meridional isomer is remarkable, because it clearly demonstrates that the morphology and thus intermolecular interactions seem to have no significant influence on the intersystem crossing process in Alq_3. However, for δ-Alq_3 there is only a very small decrease in the PL intensity and the equilibrium value remains at 98%. Due to the small decay and the noise of the measurement, the error for determination of τ_1 is too large, but from the decrease in intensity one may roughly estimate that only about 2% of the molecules in δ-Alq_3 are in the triplet state. From the similarity of the triplet states in all phases (similar zero field splitting parameters, similar lifetime) and the independence of the morphology of the samples we may conclude that the reduced population of the triplet state in δ-Alq_3 is a molecular property indicating a strongly reduced intersystem crossing in δ-Alq_3. The low concentration of the triplet states in δ-Alq_3 is consistent with the low intensity of the ODMR signal of δ-Alq_3 and the reduced intensity of its delayed fluorescence spectrum.

Discussion

Comparison of the measured properties of the triplet states in Alq_3 with theoretical calculations gives further evidence for the facial isomer in the δ-phase of Alq_3. According to the theoretical work of Amati et al. [Ama02a, Ama03] discussed in Chapter 5.1, the first S_0-S_1 transition of the facial isomer has an oscillator strength which is about a factor of 10 higher than that of the first transition of the meridional isomer (see Table 5.1). Therefore a much higher transition probability between the ground state and the first S_1-state, as well as from the S_1-state to the S_0-state, is predicted for the facial isomer. Assuming that the emission takes place from the first excited singlet state in both isomers, this predicts that the radiative emission process of the facial isomer is much faster than that of the meridional isomer. Thus, nonradiative competitive processes are less probable for the facial isomer. Therefore the probability of intersystem crossing in the facial isomer should be lower than for the meridional isomer, as observed for δ-Alq_3. From this one can conclude that the presence of the facial isomer in δ-Alq_3 is consistent with the observed low triplet concentration due to reduced intersystem crossing [Ama02a, Ama03].

The delayed luminescence spectrum in Figure 5.8 consists of two different bands with a maximum at about 525nm and 700nm, respectively. Their relative intensity is slightly different for the Alq_3-phases and dependent on temperature, as will be discussed in Ref. [Cöl04b]. The band at 525nm can be assigned to the delayed fluorescence and the band at lower energy shows all properties of the phosphorescence. But one also has to consider whether the low energy band might be related to emission from the hydroxyquinoline (8-Hq), from 8-Hq ligands that are isolated from the molecule. In order to discuss a possible contribution of 8-Hq to the delayed luminescence, the PL spectrum of polycrystalline 8-Hq was recorded using intense laser light and a wavelength of 325nm, shown together with the 700nm bands of the delayed luminescence spectra of the polycrystalline Alq_3-phases in Figure 5.13. The spectrum is very similar to that which occurs together with the delayed fluorescence, but no vibronic progression was observed in the 8-Hq-spectrum even at low temperatures (20K). 8-Hq has a very weak photoluminescence and a high excitation intensity is required. In usual PL-spectrometers the intensity is too low to result in PL from 8-Hq. This might be the reason why, to the best of our knowledge, no such spectrum for 8-Hq has been reported so far. Furthermore the 8-Hq spectrum was only observed in polycrystalline samples, not in solution, and no emission was detected using the 442nm laser line.

The lifetime of the state in 8-Hq is very short and was found by transient PL measurements to be less than 10μs, limited by the resolution of our setup. Therefore it is likely that this state is a singlet state of 8-Hq. The remarkable similarity of the spectra in Figure 5.13 supports an alternative interpretation that the observed PL spectrum is the phosphorescence of 8-Hq and the short lifetime can be explained by the dominance of nonradiative decay mechanism. This would also explain the low quantum efficiency and why the spectrum is only observed in the solid state. Further investigations have to be made to decide between these two assignments.

The vibronic progressions of the 700nm band of Alq_3 have an average distance of about $550cm^{-1}$. Vibrational modes in this energetic region are mainly related to the central Al atom (see Chapter 4.4) and the energetic distance of $550cm^{-1}$ is similar to the vibronic progressions observed for the low temperature PL measurements on Alq_3 in Chapter 5.1. Therefore the emission of the 700nm band originates from the Alq_3 molecule, and it can be excluded that the band is due to free 8-Hq molecules. This is also supported by the fact that under excitation with 442nm no emission is found from 8-Hq, but the phosphorescence band at 700nm of Alq_3 is clearly observed.

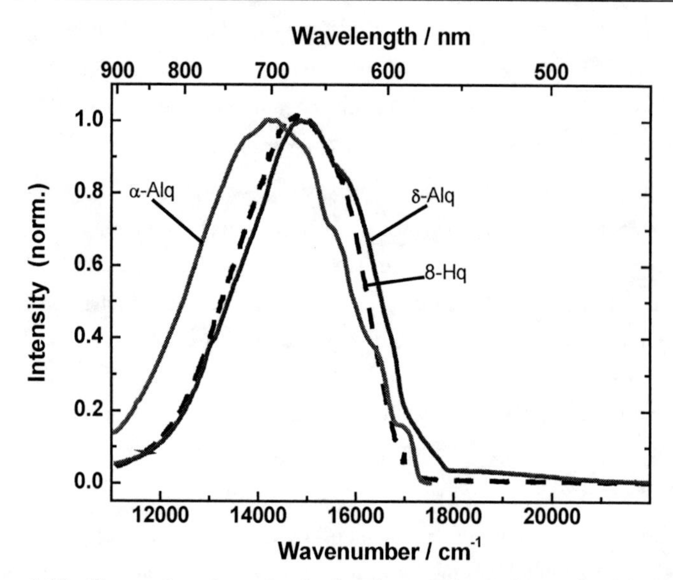

Figure 5.13: *Comparison between the bands at 700nm observed in the delayed luminescence spectrum of Alq_3 and the PL spectrum of polycrystalline hydroxyquinoline (8-Hq), which was excited by intense laser light at 325nm. The spectrum for α-Alq_3 was obtained by subtraction of its usual PL spectrum from the delayed luminescence spectrum in Figure 5.8.*

Another interpretation, namely emission from an excimer or exciplex, can also be excluded. The PL spectrum of 8-Hq in Figure 5.13 only occurs in the solid state and under intense excitation with laser light. This may suggest an emission due to the interaction of at least two 8-Hq molecules, and one might propose that in Alq_3 the ligands of neighboring molecules also overlap or interact in a similar way. However, as was shown by structural analysis in Chapter 4.3, there is a significant difference in the overlap of the ligands of neighboring Alq_3 molecules in the different phases (for example hardly any overlap in the δ-phase) [Cöl02, Bri00] and therefore the low energy band should be significantly different for the different phases. This is not the case. In fact the band at 700nm is observed in all phases and also in evaporated so-called amorphous films at approximately the same energy and with a similar shape, which indicates an intramolecular property of Alq_3.

There are two more crucial arguments for the assignment of the band at 700nm to the phosphorescence of Alq_3: First, the low energy band of the δ-phase is 30nm blue shifted, as observed for the PL spectrum and as expected for the facial isomer due to the different molecular symmetry. Second, there is always a constant factor 2 between the lifetime of the two bands (Table 5.3), as expected from theory. This was also confirmed for a wide range of temperature (Figure 5.10, [Cöl04b, Cöl04a]). Therefore it is clear that the lifetime of this band is identical with the triplet lifetime and that it is directly correlated with the triplet states in Alq_3.

From the measured delayed luminescence spectra in Figure 5.8 it is further possible to estimate the energy of the triplet states in Alq_3. As delayed fluorescence is clearly observed in all samples, the triplet energy must be more than half of the band gap and thus must be higher than 1.35eV. The appearance of the band at 700nm allows further specification of its energy. As the spectrum is the $T_1 \rightarrow S_0$ transition, one can directly determine the triplet energy from the lowest resolved vibronic band, which is probably the 0-0 transition. Hence the triplet energy for the meridional isomer can be determined as 2.11±0.1eV and for the facial isomer as 2.16±0.1eV. Martin et al. published a theoretical value of 2.13eV for the energy of the triplet state [Mar00] and Baldo et al. estimated the energy of the triplet state to be about 2eV [Bal00]. These values are in good agreement with the values experimentally determined from our data.

The results presented are the first experimental data on the triplet state of Alq_3 and they are used for comparison of the different phases of Alq_3. In addition they are the starting point for subsequent investigations in order to yield a more detailed understanding of the triplet states in Alq_3. Furthermore, this could be of interest for applications since triplet states are believed to play an important role in OLEDs.

In this chapter the excited states of Alq_3 were investigated using low temperature PL emission and PL excitation spectra as well as delayed fluorescence and ODMR measurements. For the first time the triplet state of Alq_3 was detected, its zero field splitting parameters and its lifetime were determined, and its energy was estimated. The δ-phase shows blue-shifted PL, a blue shifted excitation edge, and a strongly reduced population of the triplet state by intersystem crossing. All these differences are theoretically expected for the meridional and the facial isomer. Thus, in addition to the structural and vibrational analysis discussed in Chapter 4, the properties of the excited singlet and triplet states in Alq_3 are also in agreement with the isomerism of the molecules in the α- and δ-phase.

6 Summary

Tris(8-hydroxyquinoline)aluminum(III) (Alq_3) is a stable metal chelate complex that can be sublimed to yield amorphous thin films and represents one of the most successful organic materials used in OLEDs. It is used in a variety of display applications, for example in automotive products (e.g. Pioneer, Optrex), in mobile phones and video cameras (e.g. Kodak) or in flat panel displays (e.g. 15" OLED display presented by Sanyo). Furthermore, several theoretical models for electrical transport processes in OLEDs made of small molecular materials have been derived from measurements on Alq_3 based devices [Ber01, Brü01]. In spite of its wide usage most investigations have been carried out on evaporated amorphous thin films and comparatively few investigations have been devoted to the electronic and optical properties in the crystalline state. Furthermore, much has been speculated about the influence of the facial Alq_3 isomer in OLEDs. Even though several groups have addressed this issue, there was no experimental evidence for the presence of the facial isomer in Alq_3 and thus it was generally believed that it is thermodynamically unstable.

Therefore the questions addressed in this book were:

1. What are the properties of polycrystalline Alq_3?
2. Are there different Alq_3-phases and what are their specific characteristics?
3. Does the facial isomer exist in Alq_3 and can it be isolated?

These questions have been answered by detailed investigations of the preparation conditions and characterization of the material with a variety of different experimental methods. The samples were prepared using sublimation in a horizontal glass tube with a temperature gradient or by annealing. The thermal, structural and photophysical properties were investigated by using differential scanning calorimetry (DSC), X-ray diffraction, infrared spectroscopy, photoluminescence (PL) emission and photoluminescence excitation measurements as well as by transient PL and delayed luminescence, with the focus on the comparison of the different phases.

It was shown that different phases of Alq_3 can be separated in the sublimation tube; these differ in the shape of their crystals, in their color, their solubility, and their fluorescence. An important outcome of this work was the discovery of a new blue luminescent crystalline phase of Alq_3 (δ-Alq_3), which has significantly different properties compared to all other phases (α, β, γ). Two different crystalline phases have been identified in the sublimation tube, namely the α- and the δ-phase, and their unit cells were determined by X-ray diffraction. Furthermore, it was demonstrated that the material commonly used for the evaporation of thin films in OLEDs (here named yellowish-green Alq_3) mainly consists of α-Alq_3 with some small admixtures of δ-Alq_3.

As temperature has a strong influence on the formation of the polycrystalline phases, the formation conditions of the different phases of Alq_3 were investigated using differential scanning calorimetry measurements in combination with structural and optical characterization. As a result a phase transition at about 380°C was identified where the blue luminescent Alq_3 is formed. From detailed investigations of the processes in the temperature region between 385°C and 410°C, the two high-temperature phases δ-Alq_3 and γ-Alq_3 were identified. In addition, an efficient method was developed to prepare large amounts (several grams) of pure blue luminescent δ-Alq_3. It was also shown that all phases can easily be transformed into each other; thus chemical reactions could be excluded and the difference of the phases must be of physical and structural origin.

The well-defined preparation of pure δ-Alq_3 was the prerequisite for high resolution X-ray measurements on δ-Alq_3 powder including structural refinements. The high quality of these refinements gave convincing evidence that the facial isomer constitutes the δ-phase of Alq_3. The data also provided information about distance and orientation of the molecules and thus about molecular packing in the crystal. Compared to the α- and β-phase, both consisting of the meridional isomer, a strongly reduced π-orbital overlap of hydroxyquinoline ligands belonging to neighboring Alq_3 molecules was found in δ-Alq_3. Therefore both the packing effect with reduced intermolecular interaction and the changed symmetry of the molecule are likely to be responsible for the large blue-shift of the photoluminescence. Very recently the group at Eastman Kodak has performed structural analysis on single crystals of δ-Alq_3, prepared using the conditions as described in this book, which confirmed our findings on the structure of the facial isomer and its packing in the crystal [Tan03].

Further investigations focused on the α- and δ-phase of Alq_3. Using infrared spectroscopy it was demonstrated how the isomerism of the Alq_3 molecule is manifested in the vibrational properties. Comparison of the experimental results with theoretical calculations provided further evidence of the presence of the facial isomer in the blue luminescent δ-phase of Alq_3. In principle three aspects were discussed: First the influence of the ligands on the vibrational spectra, second the influence of the central part of the molecule, and third the effects due to the crystallinity of the measured samples. In the meridional isomer the symmetry is reduced compared to the facial isomer, leading to a higher number of bands in the spectrum, and vibrational modes of the three different ligands split into bands with slightly different energy. This results in a broadening and splitting of the observed bands in the infrared spectrum. By comparison with theoretical calculations in the literature it was shown that the measured spectra of α- and δ-Alq_3 are consistent with the presence of the meridional isomer in α-Alq_3 and of the facial isomer in δ-Alq_3. A significant indication of the different symmetry of the isomers is the first coordination sphere or central part of the molecule, which is manifested in the Al-N and Al-O stretching modes. As a result of these considerations of symmetry, characteristic features were determined in the infrared spectra, further supporting the identification of the two different isomers. Moreover, it was demonstrated that the degree of crystallinity of the samples does not influence these characteristic features.

Further important results of this work were obtained by investigations of the excited states of the different Alq_3-phases. Singlet excited states as well as triplet excited states have been investigated and from this it was possible to determine further specific properties of the different isomers.

The singlet excited states of Alq_3 were investigated using PL emission and PL excitation spectra. Both α- and δ-Alq_3 show vibronic progressions with similar energetic differences of about $550cm^{-1}$. The PL maximum of the δ-phase is blue shifted by about 0.2V and the steep excitation edge is also shifted by about 0.17eV to higher energies. These blue shifts as well as the energetic positions measured are in good agreement with theoretical calculations and thus give further evidence for the presence of the different isomers in these two phases.

This book also presents the first experiments on the triplet state of Alq_3. For the different phases of Alq_3, the zero field splitting parameters E and D were measured, the triplet lifetimes were determined, the triplet energy was estimated and the intersystem crossing behavior was investigated.

The zero field splitting parameters D and E were found to be approximately the same for all phases and only slightly dependent on the morphology (polycrystalline phases: $|D| \approx 0.063\mathrm{cm}^{-1}$, $|E| \approx 0.011\mathrm{cm}^{-1}$; evaporated film: $|D| \approx 0.061\mathrm{cm}^{-1}$, $|E| \approx 0.012\mathrm{cm}^{-1}$). Further, it was shown that the triplet lifetimes of the polycrystalline phases are approximately the same (12-16ms at 20K) and are about 60%-70% of this value for thin films. Thus the morphology seems to have only little influence on the lifetime of the triplet states, too. It was also possible to estimate the triplet energy as at least 2.11eV for the meridional isomer and 2.16eV for the facial isomer. Moreover, a new luminescence band of Alq_3 was found at about 700nm, which shows the typical characteristics of a phosphorescence spectrum. Another important result of this work was the reduced intersystem crossing in δ-Alq_3. This is in agreement with the weak intensity of the ODMR signal and the reduced intensity of the delayed fluorescence spectrum. On account of the similarity of the triplet states in all phases it was found that the reduced population of the triplet states in δ-Alq_3 due to intersystem crossing is a characteristic property of the facial isomer.

The investigations on Alq_3 presented allow the following main conclusions:

- The facial isomer exists in Alq_3 and it can be isolated in the δ-phase.
- Characteristic properties of the isomers are found in the electronic ground state as well as in the excited state of the Alq_3 molecule.

The results discussed in this book not only give conclusive and detailed answers to the questions at the beginning of this summary, but they are also the basis for further investigations in this field. So far, the properties of the facial isomer have been derived from an idealized theoretical model of the molecular structure. With the structure now measured, it is possible to specify these calculations and therefore more precise information about the properties of this isomer should be obtained in the near future. Furthermore, the facial isomer – if present in evaporated films – is believed to play an important role in OLEDs. As the facial isomer can be isolated, it is now possible to investigate its properties separately. Recently several other groups have been encouraged by the work presented here to start investigations on the δ-phase (for example at Kodak; Aixtron; Prof. Utz in USA,...). They confirm the results presented in this book [Tan03, Nan03, Utz03a]. Another important issue for the future will be to elucidate the amount of the facial isomer in evaporated films. This book also presented the first data on the triplet state of Alq_3 and opens the field for further investigations leading to a more detailed understanding of these excited states in Alq_3 as well as their role in OLEDs.

A Appendix

A.1 Synthesis of Alq_3

Tris(8-hydroxyquinoline)aluminum(III) ($Al(C_9H_6ON)_3$), commonly named Alq_3, was synthesized as follows:

15g. (100mmole) of sublimated hydroxyquinoline, 200ml of water and 35ml. of concentrated acetic acid are filled in a 1-l. three-necked flask. During continuous stirring the reaction mixture is heated up to 80°C. To the heated solution 16g. (33.7mmole) potassium aluminum sulfate dodecahydrate ($KAl(SO_4)_2*12\ H_2O$) dissolved in 150ml. of water is slowly added, followed by an additional 20ml. (40mmole) of hydrochloric acid, and a yellowish precipitation is observed. Then a solution of 100ml. (200mmole) of ammonium acetate is added drop by drop. After the addition is completed, the reaction mixture is heated up to boiling point for a short time and is then allowed to cool down to room temperature over night under continuous stirring. Subsequently the solvent is removed under reduced pressure, and the residue is washed with water and then methanol until pH 5-6 is reached. Finally the product is dried under vacuum at 60°C for several hours and tris(8-hydroxyquinoline)aluminum(III) ($Al(C_9H_6ON)_3$) (Alq_3) is obtained (14g.).

A.2 Supplementary Data for the Results Discussed in Chapter 4

Table A.1: *Observed and calculated peaks from the X-ray powder diffractogram of blue Alq_3 (Chapter 4) in the 2Θ range 5-32° together with the indexing*[4].

Peak No.	2Θ(°) (obs)	H	K	L	2Θ(°) (calc.)	Intensity (a.u.)
1	6.69	0	0	1	6.69	78
2	7.32	0	1	0	7.32	86
3	7.66	0	-1	1	7.65	100
4	11.76	0	1	1	11.77	18
5	13.41	0	0	2	13.42	28
6	14.40	1	0	0	14.40	45
7	14.70	0	2	0	14.68	6
8	15.49	-1	1	0	15.49	89
9	15.79	-1	0	1	15.79	48
10	16.00	1	0	1	16.01	63
11	16.86	-1	-1	1	16.88	30
12	17.74	0	1	2	17.75	16
13	18.46	0	-1	3	18.46	43
14	18.71	1	-2	1	18.70	5
15	19.31	1	1	1	19.31	13
16	19.55	-1	0	2	19.56	32
17	20.31	0	-3	1	20.30	12
18	20.73	-1	-2	1	20.73	3
19	21.65	1	2	0	21.66	14
20	22.09	0	3	0	22.10	3
21	23.58	1	1	2	23.56	58
22	23.69	1	-2	3	23.69	77
23	25.08	-1	-2	3	25.06	45
24	26.90	-1	2	2	26.89	4
25	28.33	-1	3	1	28.32	5
26	28.76	1	2	2	28.75	14
27	29.04	2	0	0	29.05	12
28	29.23	-2	1	0	29.24	11
29	29.45	2	-1	1	29.45	25
30	29.76	1	-4	1	29.80	10
31	30.70	-2	-1	1	30.70	5
32	31.70	-1	-3	4	31.69	9

[4] In the first report on blue luminescent Alq_3 (fraction 1) in Reference [Bra01] we included the shoulder at 7.05°, which led to a different unit cell. As shown in Chapter 4.2 this shoulder is related to the γ-phase.

Table A.2: *Observed and calculated peaks from the X-ray powder diffractogram of fraction 3 (α-Alq_3) in Chapter 4 together with the indexing.*

Peak No.	**2Θ(°) (obs)**	**H**	**K**	**L**	**2Θ(°) (calc.)**	**Intensity (a.u.)**
1	6.32	0	1	0	6.37	42
2	7.31	1	0	0	7.34	42
3	7.84	-1	1	0	7.92	100
4	11.23	1	1	0	11.25	24
5	12.74	0	2	0	12.76	28
6	13.85	-2	1	0	13.90	6
7	14.24	0	0	1	14.27	16
8	15.11	-1	0	1	15.12	42
9	15.41	0	-1	1	15.40	21
10	15.83	-2	2	0	15.87	52
11	16.92	1	0	1	16.97	14
12	17.98	2	1	0	17.94	30
13	19.17	0	3	0	19.18	7
14	19.74	-2	3	0	19.84	5
15	20.84	-3	1	0	20.84	1
16	21.94	2	0	1	21.96	3
17	23.12	1	2	1	23.11	46
18	23.67	-3	1	1	23.67	20
19	24.49	0	3	1	24.49	14
20	24.70	-3	0	1	24.67	16
21	25.25	2	-3	1	25.26	17
22	27.32	3	-2	1	27.30	1
23	28.19	1	-4	1	28.15	9
24	29.23	3	2	0	29.24	6

Supplementary data from the determination of the molecular structure using radiation from a synchrotron light source

A cif-file of these data is available at http://www.rsc.org/suppdata/cc/b2/b209164j/

Table A.3: *Positional parameters of the facial Alq_3 molecule in δ-Alq_3.*

Name	X	Y	Z	**Name**	X	Y	Z
Al1	0.6865(6)	0.3116(5)	0.774(1)	**C18**	1.0032(8)	0.1877(8)	0.697(2)
N2	0.5863(7)	0.1964(7)	0.545(2)	**C19**	0.9561(9)	0.1288(8)	0.878(2)
C3	0.6088(7)	0.1556(8)	0.370(2)	**C20**	0.8629(9)	0.1384(7)	0.965(2)
C4	0.5217(9)	0.0790(9)	0.242(1)	**C21**	0.8207(7)	0.2079(7)	0.866(1)
C5	0.4110(9)	0.0416(7)	0.295(2)	**C22**	0.8679(6)	0.2668(6)	0.685(1)
C6	0.3879(6)	0.0828(7)	0.479(1)	**O23**	0.7366(8)	0.2234(7)	0.938(1)
C7	0.2770(7)	0.0461(7)	0.546(2)	**N24**	0.6499(7)	0.4164(6)	0.553(2)
C8	0.2598(7)	0.0918(8)	0.732(2)	**C25**	0.5883(8)	0.4033(7)	0.357(2)
C9	0.3520(8)	0.1728(9)	0.853(2)	**C26**	0.5713(8)	0.4857(9)	0.255(1)
C10	0.4617(7)	0.2064(6)	0.788(1)	**C27**	0.6189(7)	0.5820(8)	0.353(2)
C11	0.4802(6)	0.1615(6)	0.603(1)	**C28**	0.6800(7)	0.5950(6)	0.560(1)
O12	0.5513(7)	0.2799(7)	0.893(1)	**C29**	0.7286(8)	0.6917(7)	0.673(2)
N13	0.8224(6)	0.3310(7)	0.592(2)	**C30**	0.7908(8)	0.7017(7)	0.877(2)
C14	0.8662(9)	0.3888(8)	0.422(2)	**C31**	0.8053(7)	0.6147(8)	0.973(2)
C15	0.958(1)	0.3824(7)	0.334(2)	**C32**	0.7559(7)	0.5181(6)	0.859(1)
C16	1.0035(7)	0.3156(8)	0.422(2)	**C33**	0.6936(6)	0.5082(6)	0.655(1)
C17	0.9585(7)	0.2553(7)	0.600(1)	**O34**	0.7639(7)	0.4326(6)	0.936(1)

Table A.4: *Bond lengths obtained for the refined structure.*

Bond	**Distance**	**Bond**	**Distance**	**Bond**	**Distance**
Al1_N2	2.112(9)	**O34_C32**	1.351(7)	**C17_C22**	1.429(7)
Al1_N13	2.14(1)	**C3_C4**	1.411(8)	**C18_C19**	1.425(8)
Al1_N24	2.18(1)	**C4_C5**	1.413(7)	**C19_C20**	1.451(8)
Al1_O12	1.883(7)	**C5_C6**	1.407(8)	**C20_C21**	1.430(7)
Al1_O23	1.890(7)	**C6_C11**	1.450(7)	**C21_C22**	1.424(7)
Al1_O34	1.877(7)	**C6_C7**	1.446(7)	**C25_C26**	1.418(8)
N2_C11	1.370(7)	**C7_C8**	1.423(8)	**C26_C27**	1.396(8)
N2_C3	1.362(8)	**C8_C9**	1.455(8)	**C27_C28**	1.416(7)
N13_C14	1.351(8)	**C9_C10**	1.430(7)	**C28_C29**	1.440(8)
N13_C22	1.381(7)	**C10_C11**	1.418(7)	**C28_C33**	1.438(7)
N24_C25	1.362(7)	**C14_C15**	1.418(8)	**C29_C30**	1.408(8)
N24_C33	1.353(7)	**C15_C16**	1.402(8)	**C30_C31**	1.449(8)
O12_C10	1.343(7)	**C16_C17**	1.410(8)	**C31_C32**	1.442(7)
O23_C21	1.339(7)	**C17_C18**	1.425(8)	**C32_C33**	1.409(7)

Table A.5: *Selected angles obtained from the Rietveld refinement of the facial isomer of δ-Alq_3.*

Angle	Degree	Angle	Degree	Angle	Degree	Angle	Degree
Al1_N13_C14	134.2(7)	**C8_C9_C10**	120.2(6)	**C25_N24_C33**	120.1(6)	**N2_C11_C6**	121.8(5)
Al1_N13_C22	105.2(6)	**C9_C10_C11**	119.8(5)	**C26_C27_C28**	119.0(6)	**N2_C3_C4**	119.8(6)
Al1_N2_C11	106.6(6)	**C9_C10_O12**	123.8(6)	**C27_C28_C29**	121.6(7)	**N13_Al1_N24**	87.8(4)
Al1_N2_C3	133.0(7)	**C11_C10_O12**	116.4(6)	**C27_C28_C33**	117.8(6)	**N13_Al1_O23**	83.1(5)
Al1_N24_C25	132.8(8)	**C14_C15_C16**	120.3(7)	**C28_C29_C30**	119.8(6)	**N13_Al1_O34**	94.2(5)
Al1_N24_C33	106.9(6)	**C14_N13_C22**	120.6(6)	**C28_C33_C32**	119.7(6)	**N13_C14_C15**	120.2(7)
Al1_O12_C10	115.8(7)	**C15_C16_C17**	119.8(6)	**C29_C28_C33**	120.6(6)	**N13_C22_C17**	121.5(5)
Al1_O23_C21	116.5(7)	**C16_C17_C18**	121.8(6)	**C29_C30_C31**	120.0(7)	**N13_C22_C21**	119.1(6)
Al1_O34_C32	118.3(7)	**C16_C17_C22**	117.6(6)	**C30_C31_C32**	119.8(6)	**N24_Al1_O34**	80.7(5)
C3_C4_C5	121.5(7)	**C17_C18_C19**	120.1(6)	**C31_C32_C33**	120.2(6)	**N24_C25_C26**	120.7(7)
C3_N2_C11	120.5(6)	**C17_C22_C21**	119.3(6)	**C31_C32_O34**	124.2(6)	**N24_C33_C28**	122.1(5)
C4_C5_C6	118.7(6)	**C18_C17_C22**	120.6(6)	**C33_C32_O34**	115.7(6)	**N24_C33_C32**	118.2(6)
C5_C6_C11	117.7(5)	**C18_C19_C20**	120.0(7)	**N2_Al1_N13**	86.9(4)	**O12_Al1_N13**	170.0(6)
C5_C6_C7	121.6(6)	**C19_C20_C21**	118.8(6)	**N2_Al1_N24**	86.4(4)	**O12_Al1_N24**	89.9(5)
C6_C11_C10	120.1(5)	**C20_C21_C22**	121.2(6)	**N2_Al1_O12**	83.2(5)	**O12_Al1_O23**	99.1(6)
C6_C7_C8	118.5(6)	**C20_C21_O23**	123.0(6)	**N2_Al1_O23**	92.8(5)	**O12_Al1_O34**	95.1(6)
C7_C6_C11	120.7(6)	**C22_C21_O23**	115.8(6)	**N2_Al1_O34**	167.0(6)	**O23_Al1_N24**	170.9(6)
C7_C8_C9	120.7(6)	**C25_C26_C27**	120.3(6)	**N2_C11_C10**	118.0(6)	**O23_Al1_O34**	100.2(6)

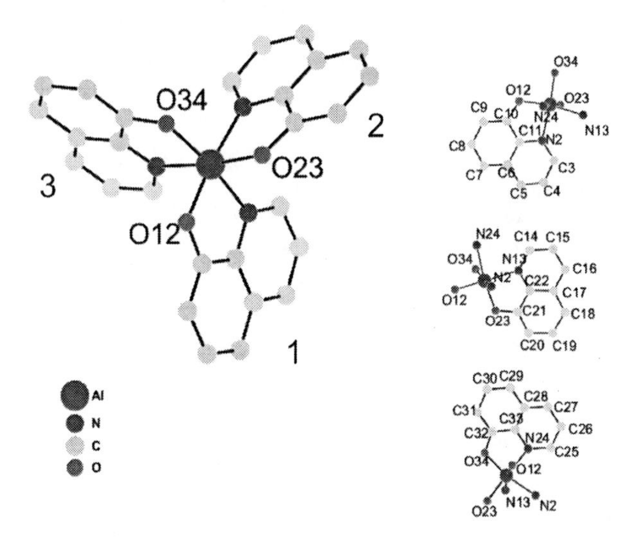

Figure A.1: *Labeling of the atoms for the molecular structure of the facial Alq_3 molecule.*

A.3 Obtained Results on the Assumption of the Meridional Isomer

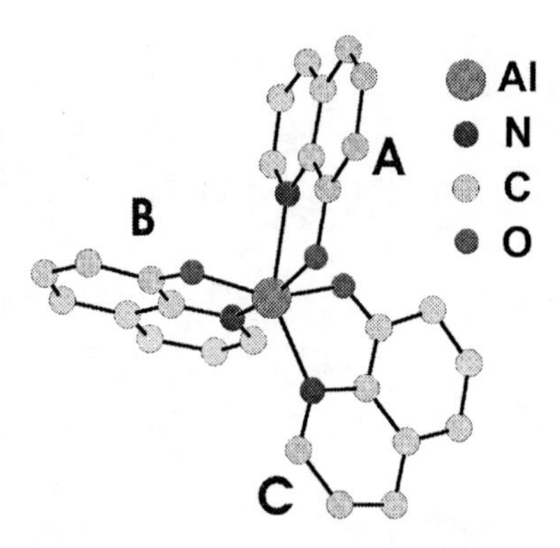

Table A.6:*Crystallographic data for δ-Alq_3. R_p, R_{wp} and R-F^2 refer to the Rietveld criteria of fit for profile and weighted profile respectively, defined by Langford and Louer [Lan96]*

Formula	Al(C9H6NO)3
Temperature [K]	295
Formula weight [g/mol]	918.88
Space group	P-1
Z	2
a [Å]	13.2412(1)
b [Å]	14.4245(1)
c [Å]	6.17684(4)
α [°]	88.5553(7)
β [°]	95.9230(6)
γ [°]	113.9332(5)
V [$Å^3$]	1072.38(2)
ρ-calc [g/cm^3]	1.423
2Θ range [°]	4-35.7
Step size [°2Θ]	0.005
Wavelength [Å]	1.14982(2)
μ [1/cm]	2.48
Capillary diamater	0.7
R-p [%]	7.3
R-wp [%]	9.4
R-F^2 [%]	19.4
Reduced χ^2	3.557
No of reflections	357

Table A.7: *Positional parameters and temperature factors for δ-Alq_3 at ambient conditions for the assumption of the meridional isomer*

Name	X	Y	Z	Ui Å²*100	Name	X	Y	Z	Ui Å²*100
AL1	0.182(1)	0.3033(8)	0.806(2)	2.4(5)	**C18**	-0.218(1)	0.047(1)	0.527(3)	1.3(2)
N2	0.149(1)	0.4182(9)	0.553(3)	1.3(2)	**C19**	-0.237(1)	0.093(1)	0.702(3)	1.3(2)
C3	0.089(1)	0.410(1)	0.348(3)	1.3(2)	**C20**	-0.148(1)	0.171(1)	0.832(2)	1.3(2)
C4	0.071(1)	0.491(1)	0.253(2)	1.3(2)	**C21**	-0.037(1)	0.204(1)	0.783(2)	1.3(2)
C5	0.121(1)	0.583(1)	0.364(3)	1.3(2)	**C22**	-0.015(1)	0.1628(9)	0.605(2)	1.3(2)
C6	0.181(1)	0.5965(9)	0.571(2)	1.3(2)	**O23**	0.046(1)	0.274(1)	0.899(2)	1.3(2)
C7	0.227(1)	0.694(1)	0.690(3)	1.3(2)	**N24**	0.291(1)	0.2304(1)	0.869(2)	1.3(2)
C8	0.287(1)	0.705(1)	0.896(3)	1.3(2)	**C25**	0.300(1)	0.157(1)	0.990(3)	1.3(2)
C9	0.305(1)	0.619(1)	0.994(2)	1.3(2)	**C26**	0.388(2)	0.123(1)	0.971(2)	1.3(2)
C10	0.257(1)	0.5168(9)	0.879(2)	1.3(2)	**C27**	0.465(1)	0.152(1)	0.817(3)	1.3(2)
C11	0.192(1)	0.5092(9)	0.666(2)	1.3(2)	**C28**	0.451(1)	0.227(1)	0.682(2)	1.3(2)
O12	0.266(1)	0.4267(9)	0.955(2)	1.3(2)	**C29**	0.519(1)	0.272(1)	0.519(3)	1.3(2)
N13	0.091(1)	0.198(1)	0.558(2)	1.3(2)	**C30**	0.499(1)	0.344(1)	0.370(2)	1.3(2)
C14	0.113(1)	0.160(1)	0.377(3)	1.3(2)	**C31**	0.421(1)	0.382(1)	0.390(2)	1.3(2)
C15	0.033(1)	0.086(1)	0.245(2)	1.3(2)	**C32**	0.352(1)	0.330(1)	0.549(2)	1.3(2)
C16	-0.079(1)	0.046(1)	0.291(2)	1.3(2)	**C33**	0.367(1)	0.2624(9)	0.706(2)	1.3(2)
C17	-0.105(1)	0.084(1)	0.471(2)	1.3(2)	**O34**	0.281(1)	0.358(1)	0.595(2)	1.3(2)

Table A.8: *Bondlengths [Å] for δ-Alq_3 at ambient conditions for the assumption of the meridional isomer.*

Bond	Distance	Bond	Distance	Bond	Distance
AL1_N13	2.09(1)	**C17_C22**	1.46(1)	**C3_C4**	1.39(1)
AL1_N2	2.39(2)	**C18_C19**	1.38(1)	**C3_N2**	1.40(1)
AL1_N24	2.10(1)	**C19_C20**	1.45(1)	**C30_C31**	1.37(1)
AL1_O12	1.87(1)	**C20_C21**	1.41(1)	**C31_C32**	1.40(1)
AL1_O23	1.84(1)	**C21_C22**	1.38(1)	**C32_C33**	1.42(1)
AL1_O34	1.87(1)	**C21_O23**	1.31(1)	**C32_O34**	1.22(1)
C10_C11	1.47(1)	**C22_N13**	1.34(1)	**C33_N24**	1.42(1)
C10_O12	1.42(1)	**C25_C26**	1.45(1)	**C33_O34**	2.17(1)
C11_N2	1.38(1)	**C25_N24**	1.32(1)	**C4_C5**	1.38(1)
C14_C15	1.37(1)	**C26_C27**	1.40(1)	**C5_C6**	1.40(1)
C14_N13	1.37(1)	**C27_C28**	1.41(1)	**C6_C11**	1.44(1)
C15_C16	1.42(1)	**C28_C29**	1.39(1)	**C6_C7**	1.46(1)
C16_C17	1.38(1)	**C28_C33**	1.42(1)	**C7_C8**	1.41(1)
C17_C18	1.45(1)	**C29_C30**	1.46(1)	**C8_C9**	1.47(1)
				C9_C10	1.51(1)

Table A.9: *Angles [°] for δ-Alq_3 at ambient conditions for the assumption of the meridional isomer:*

Angle	Degrees	Angle	Degrees	Angle	Degrees
O12_AL1_N13	161.2(9)	**C11_C10_O12**	116.3(9)	**C17_C22_C21**	120.5(8)
O12_AL1_O23	100.2(8)	**N2_C11_C6**	120.0(8)	**AL1_O23_C21**	115(1)
O12_AL1_N24	100.1(8)	**N2_C11_C10**	119.0(9)	**AL1_N24_C25**	139(1)
O12_AL1_O34	84.5(8)	**C6_C11_C10**	120.9(9)	**AL1_N24_C33**	107(1)
N13_AL1_O23	83.5(8)	**AL1_O12_C10**	120(1)	**C25_N24_C33**	113.7(9)
N13_AL1_N24	91.1(7)	**AL1_N13_C14**	136(1)	**N24_C25_C26**	121(1)
N13_AL1_O34	83.1(7)	**AL1_N13_C22**	105.6(9)	**C25_C26_C27**	127(1)
O23_AL1_N24	130.1(8)	**C14_N13_C22**	118.1(9)	**C26_C27_C28**	110.4(9)
O23_AL1_O34	148.9(9)	**N13_C14_C15**	123(1)	**C27_C28_C29**	122(1)
N24_AL1_O34	78.1(7)	**C14_C15_C16**	120(1)	**C27_C28_C33**	121.7(8)
C3_N2_C11	119.7(9)	**C15_C16_C17**	118.2(9)	**C29_C28_C33**	115.9(9)
N2_C3_C4	122(1)	**C16_C17_C18**	121.3(9)	**C28_C29_C30**	121.7(9)
C3_C4_C5	116.8(9)	**C16_C17_C22**	118.5(8)	**C29_C30_C31**	124(1)
C4_C5_C6	124.1(9)	**C18_C17_C22**	120.2(8)	**C30_C31_C32**	110.9(9)
C5_C6_C7	123(1)	**C17_C18_C19**	117.5(9)	**C31_C32_C33**	128.1(8)
C5_C6_C11	116.9(9)	**C18_C19_C20**	121.7(9)	**C31_C32_O34**	119(1)
C7_C6_C11	120.3(8)	**C19_C20_C21**	120.8(9)	**C33_C32_O34**	111.0(9)
C6_C7_C8	121.5(9)	**C20_C21_C22**	119.1(8)	**N24_C33_C28**	125.1(8)
C7_C8_C9	120(1)	**C20_C21_O23**	123(1)	**N24_C33_C32**	116.9(9)
C8_C9_C10	120.3(9)	**C22_C21_O23**	118.2(9)	**C28_C33_C32**	117.9(8)
C9_C10_C11	117.2(8)	**N13_C22_C17**	121.9(8)		
C9_C10_O12	126.5(9)	**N13_C22_C21**	117.6(9)		

References

[Ada01] Adachi, C., Baldo, M., Thompson, M., and Forrest, S., "Nearly 100% internal phosphorescence efficiency in an organic light emitting device", *J. Appl. Phys*. 90, 5048 (2001).

[Alb95a] Albrecht, U., and Bässler, H., "Efficiency of charge recombination in organic light emitting diodes", *Chem. Phys.* 199, 207 (1995).

[Alb95b] Albrecht, U., and Bässler, H., "Langevin-type charge carrier recombination in a disordered hopping system", *Phys. Status Solidi B* 191, 455 (1995).

[Ama02] Amati, M., and Lelj, F., "Monomolecular isomerization processes of aluminum tris(8-hydroxyquinolinate) (Alq_3): a DFT study of gas-phase reaction paths", *Chem. Phys. Lett.* 363, 451 (2002).

[Ama02a] Amati, M., and Lelj, F., "Are uv-vis and luminescence spectra of Alq_3 (aluminum tris(8-hydroxyquinolinate)) delta-phase compatible with the presence of the fac-Alq_3 isomer? A TD-DFT investigation", *Chem. Phys. Lett.* 358, 144 (2002).

[Ama03] Amati, M., and Lelj, F., "Luminescent compounds fac- and mer- aluminum tris(quinolin-8-olate). A pure and hybrid density functional theory and time-dependent density functional theory investigation of their electronic and spectroscopic properties", *J. Phys. Chem. A.* 107, 2560 (2003).

[And98] Anderson, J., McDonald, E., Lee, P., Anderson, M., Ritchie, E., Hall, H., Hopkins, T., Mash, E., Wang, J., Padias, A., Thayumanavan, S., Barlow, S., Marder, S., Jabbour, G., Shaheen, S., Kippelen, B., Peyghambarian, N., Wightman, R., and Arm., N., "Electrochemistry and electrogenerated chemiluminescence processes of the components of aluminum quinolate/triarylamine, and related organic light-emitting diodes", *J. Am. Chem. Soc.* 120, 9646 (1998).

[Azi98] Aziz, H., Popovic, Z., Tripp, C., Hu, N. X., Hor, A., and Xu, G., "Degradation processes at the cathode/organic interface in organic light emitting devices with Mg:Ag cathodes", *Appl. Phys. Lett.* 72, 2642 (1998).

[Azi99] Aziz, H., Popovic, Z., Hu, N.-X., Hor, A.-M., and Xu, G., "Degradation mechanism of small molecule-based organic light-emitting devices", *Science* 283, 1900 (1999).

[Bak68] Baker, B. C., and Sawyer, D. T., "Proton nuclear magnetic resonance studies of 8-quinolinol and several of its metal complexes", *Anal. Chem.* 40, 1945 (1968).

[Bal00] Baldo, M., and Forrest, S., "Transient analysis of organic electrophosphorescence: Transient analysis of triplet energy transfer", *Phys. Rev. B* 62, 10958 (2000).

[Bal99] Baldo, M., Lamansky, S., Burrows, P.E., Thompson, M., and Forrest, S., "Very high-efficiency green organic light-emitting devices based on electrophosphorescence", *Appl. Phys. Lett.* 75, 4 (1999).

[Bar99] Barth, S., Wolf, U., Bässler, H., Müller, P., Riel, H., Vestweber, H., Seidler, P., and Rieß, W., "Current injection from a metal to a disordered hopping system. III. comparison between experiment and monte carlo simulation", *Phys. Rev. B* 60, 8791 (1999).

[Ber01] Berleb, S. (2001). *Raumladungsbegrenzte Ströme und Hoppingtransport in organischen Leuchtdioden aus Tris-(8-hydroxyquinolin)-Aluminium (Alq_3).* PhD thesis, University of Bayreuth.

[Bor93] Borsenberger, D.S., and Weiss, P. M. (1993). *Organic Photoreceptors for Imaging Systems*. M. Dekker, New York.

[Bra99] Braun, M. (1999). *Selbstorganisierte organische Schichten in Bleihalogenid-Kristallen.* PhD thesis, University of Stuttgart.

[Bra01] Braun, M., J.Gmeiner, M.Tzolov, Cölle, M., Meyer, F., W.Milius, H.Hillebrecht, Wendland, O., von Schütz, J., and Brütting, W., "A new crystalline phase of the electroluminescent material tris(8-hydroxyquinoline) aluminum exhibiting blue-shifted fluorescence", *J. Chem. Phys.* 114, 9625 (2001).

[Bri00] Brinkmann, M., Gadret, G., Muccini, M., Taliani, C., Masciocchi, N., and Sironi, A., "Correlation between molecular packing and optical properties in different crystalline polymorphs and amorphous thin films of mer-tris(8-hydroxyquinoline)aluminium(III)", *J. Am. Chem. Soc.* 122, 5147 (2000).

[Bro73] Brown, D., "Chemistry of vitamin B_{12} and related inorganic model systems", *Prog. Inorg. Chem.* 18, 177 (1973).

[Brü01] Brütting, W., Berleb, S., and Mückl, A., "Device physics of organic lightemitting diodes based on molecular materials", *Organic Electronics* 2, 1 (2001).

[Bur96] Burrows, P., Shen, Z., Bulovic, V., McCarty, D., Forrest, S., Cronin, J., and Thompson, M., "Relationship between electroluminescence and current transport in organic heterojunction LED's", *J. Appl. Phys.* 79, 7991 (1996).

[Cöl02] Cölle, M., Dinnebier, R. E., and Brütting, W., "The structure of the blue luminescent delta-phase of tris(8-hydroxyquinoline)alminium(III) Alq_3", *Chem. Comm.* 23, 2908 (2002).

[Cöl02a] Cölle, M., Gmeiner, J., Milius, W., Hillebrecht, H., and Brütting, W., "Thermal and structural properties of the organic electroluminescent material tris(8-hydroxyquinoline)aluminum (Alq_3)", *ProceedingsEL2002*, 133, (2002).

[Cöl03] Cölle, M., Gmeiner, J., Milius, W., Hillebrecht, H., and Brütting, W., "Preparation and characterization of blue-luminescent tris(8-hydroxyquinoline)aluminum (Alq_3)", *Adv. Funct. Mater.* 13, 108 (2003).

[Cöl03a] Cölle, M., Forero-Lenger, S., Gmeiner, J., and Brütting, W., "Vibrational analysis of different crystalline phases of the organic electroluminescent material aluminium tris(quinoline-8-olate) (Alq_3).", *Phys. Chem. Chem. Phys.* 5, 2958 (2003).

[Cöl04] Cölle, M., and Brütting, W., "Thermal, structural and photophysical properties of Alq_3", *Phys. stat. sol. A,* in press.

[Cöl04a] Cölle, M., and Gärditz, C., "Phosphorescence of Alq_3", submitted.

[Cöl04b] Cölle, M., Gärditz, C.,and Braun, M., "The triplet state of Alq_3", submitted.

[Cha67] Chalmers, R., and Basit, M., "A critical study of 8-hydroxyquinoline as a gravimetric reagent for aluminium", *The Analyst* 92, 680 (1967).

[Che98] Chen, C., and Shi, J., "Metal chelates as emitting materials for organic electroluminescence", *Coord. Chem. Rev.* 171, 161 (1998).

[Cho99] Choong, V., Shi, S., Curles, J., Shieh, C., Lee, H., So, F., Shen, J., and Yang, J., "Organic light-emitting diodes with a bipolar transport layer", *Appl. Phys. Lett.* 75, 172 (1999)

[Cla82] Clarke, R.H. (1982). *Triplet State ODMR Spectroscopy*. Wiley, New York.

[Cop92] Coppens, P. (1992). *Synchrotron Radiation Crystallography*. Academic Press, London.

[Cur02] Curry, R.J., Gillin, W.P., Clarkson, J., and Batchelder, D.N., "Morphological study of aluminum tris(8-hydroxyquinoline) thin films using infrared and Raman spectroscopy", *J. Appl. Phys.* 92, 1902 (2002).

[Cur98] Curioni, A., Boero, M., and Andreoni, W., "Alq_3: ab initio calculations of its structural and electronic properties in neutral and charged states", *Chem. Phys. Lett.* 294, 263 (1998).

[Esp02] Degli Esposti, A., Brinkmann, M., and Ruani, G., "The dynamics of the internal phonons tris(quinolin-8-olato) aluminum (III) in crystalline beta-phase", *J. Chem. Phys.* 116, 798 (2002).

[For89] Ford, J., and Timmins, P., (1989). *Pharmaceutical Thermal Analysis*.

[For98] Forsythe, E., Morton, D., Tang, C., and Gao, Y., "Trap states of tris-8-(hydroxyquinoline) aluminum and naphthyl-substituted benzidine derivative using thermally stimulated luminescence", *Appl. Phys. Lett.* 73, 1457 (1998).

[Fuj96] Fujii, I., Hirayama, N., Ohtani, J., and Kodama, K., "Crystal structure of tris(8-quinolinolato)alumium(III)-ethyl acetate(1/0.5)", *Analytical Sciences* 12, 153 (1996).

[Gär03] Gärditz, C. (2003). Diplomarbeit. University of Bayreuth.

[Gar96] Garbuzov, D., Bulovic, B., Burrows, P., and Forrest, S., "Photoluminescence efficiency and absorption of aluminum-tis-quinolate (Alq_3) thin films", *Chem. Phys. Lett.* 249, 433 (1996).

[Gee85] Geels, F., Schmidt, E., and Schippers, B., "The use of 8-hydroxyquinoline for the isolation and prequalification of plant growing rhizpsphere pseudomonads", *Biology and Fertility of Soils* 1, 167 (1985).

[Hal63] Hall, J., Jennings, D., and McClintock, R., "Study of anthracene fluorescence excited by the ruby giant–pulse laser", *Phys. Rev. Lett.* 11, 364 (1963).

[Hal98] Halls, M., and Aroca, R., "Vibrational spectra and structure of tris(8-hydroxyquinoline) aluminium(III)", *Can. J. Chem.* 76, 1730 (1998).

[Ham93] Hamada, Y., Sano, T., Fujita, M., Fujii, T., Nishio, Y., and Shibata, K., "Organic electroluminescent devices with 8-hydroxyquinoline derivative-metal complexes as an emitter", *Jpn. J. Appl. Phys.* 32, L514 (1993).

[Ham97] Hamada, Y., Sano, T., Fujii, H., Nishio, Y., Takahashi, H., and Shibata, K., "Organic light-emitting diodes using 3- or 5-hydroxyflavone metal complexes", *Appl. Phys. Lett.* 71, 3338 (1997).

[Hil90] Hill, R.J., and Fischer, R.X., "Profile agreement indices in Riėtveld and pattern-finding analysis", *J. Appl. Cryst.* 23, 462 (1990).

[Hum00] Humbs, W., Zhang, H., and Glasbeek, M., "Femtosecond fluorescence upconversion spectroscopy of vapor-deposited tris(8-hydroxyquinoline) aluminum films", *Chem. Phys.* 254, 319 (2000).

[Hun97] Hung, L., Tang, C., and Mason, M., "Enhanced electron injection in organic electroluminescence devices using an Al/LiF electrode", *Appl. Phys. Lett.* 70, 152 (1997).

[Ich01] Ichikawa, H., Shimada, T., and Koma, A., "Ordered growth and crystal structure of Alq_3 on alkali halide surfaces", *Jpn. J. Appl. Phys. 2* 40, 225 (2001).

[Ich02] Ichikawa, M., Yanagi, H., Shimizu, Y., Hotta, S., Suganuma, N., Koyama, T., and Taniguchi, Y., "Organic field-effect transistors made of epitaxially grown crystals of a thiophene/phenylene co-oligmer", *Adv. Mater.* 14, 1272 (2002).

[Ito02] Ito, E., Washizu, Y., Hayashi, N., Ishii, H., Matsuie, N., Tsuboi, K., Ouchi, Y., Harima, Y., Yamashita, K., and Seki, K., "Spontaneous buildup of giant surface potential by vacuum deposition of Alq_3 and its removal by visible light irradiation", *J. Appl. Phys.* 92, 7306 (2002).

[Jeo02] Jeong, Y., Troadec, D., Moliton, A., Ratier, B., Antony, R., and Veriot, G., "Dielectric studies of Alq_3 and transport mechanisms", *Synth. Met.* 127, 195 (2002).

[Joh99] Johansson, N., Osada, T., Stafström, S., Salaneck, W., Parente, V., dos Santos, D., Crispin, X., and Brédas, J., "Electronic structure of tris(8-hydroxyquinoline)aluminum thin films in the pristine and reduced states", *J. Chem. Phys.* 111, 2157 (1999).

[Jon00] Jonda, C., Mayer, A., Stolz, U., Elschner, A., and Karbach, A., "Surface roughness effects and their influence on the degradation of organic light emitting devices", *J. Mat. Science* 35, 5645 (2000).

[Kao81] Kao, W., and Hwang, K. (1981). *Electrical Transport in Solids*. Pergamon Press, Oxford.

[Kau74] Kauffmann, G., "Alfred Werner's research on optically active coordination compounds", *Coord. Chem. Rev.* 12, 105 (1974).

[Kaw01] Kawasumi, Y., Akai, I., and Karasawa, T., "Photoluminescence and dynamics of excitons in Alq_3 single crystals", *Int. J. Mod. Phys. B* 15, 3825 (2001).

[Kep95] Kepler, R., Beeson, P., Jacobs, S., Anderson, R., Sinclair, M., Valencia, V., and Cahill, P., "Electron and hole mobility in tris(8-hydroxyquinolinolatio) aluminium", *Appl. Phys. Lett.* 66, 3618 (1995).

[Kim95] Kim, S., and Karis, T., "Glass formation from low molecular weight organic melts", *J. Mater. Res.* 10, 2128 (1995).

[Kis03] Kishore, V. R., Narasimhan, K., and Periassamy, N., "On the radiative lifetime, quantum yield and fluorescence decay of Alq_3 in thin films", *Phys. Chem. Chem. Phys.* 5, 1386 (2003).

[Kub00] Kubota, H., Miyaguchi, S., Ishizuka, S., Wakimoto, T., Funaki, J., Fukuda, Y., Watanabe, T., Ochi, H., Sakamoto, T., Miyake, T., Tsuchida, M., Ohshita, I., and Tohma, T., "Organic LED full color passive-matrix display", *J. Lumin.* 87-89, 56 (2000).

[Kus00] Kushto, G., Iizumi, Y., Kido, J., and Kafafi, Z., "A matrixisolation spectroscopic and theoretical investigation of tris(8-hydroxyquinolinato)aluminum(III) and tris(4-methyl-8-hydroxyquinolinato) aluminum(III)", *J. Phys. Chem. A* 104, 3670 (2000).

[Lan96] Langford, I., and Louer, D., "Powder diffraction", *Rep. Prog. Phys.* 59, 131 (1996).

[Lar68] Larsson, R., and Eskilsson, O., "On the far infrared spectra of the trisoxinato complexes of Al(III), Fe(III), and Co(III) in chloroform solution", *Acta Chem. Scand.* 22, 1067 (1968).

[Leh91] Lehmann, H.D., "Treatment of Alzheimer's disease with 8-hydroxyquinoline and 8-hydoxyquinaldine derivatives", Patent No. DE 3932338, (1991).

[Maj70] Majer, J., and Reade, M., "Isomerism in the metal derivatives of 8-hydroxyquinoline", *Chem. Comm.* 1, 58 (1970).

[Mar00] Martin, R., Kress, J., Campbell, I., and Smith, D., "Molecular and solid-state properties of tris-(8-hydroxyquinolate)-aluminium", *Phys. Rev. B* 61, 15804 (2000).

[Mar01] Markham, P.M., Klyachko, E.A., Crich, D., Jaber, M-R., Johnson, M., Mulhearn,D.C., Neyfakh, A.A., "Bactericidal methods and compositions using acyl hydrazides, oxamides and 8-hydroxyquinolines as antibiotic potentiators for treatment of Gram-positive infections", Patent No. WO2001070213, 2001-US9578 (2001)

[Mas00] Massa, W. (2000). *Crystal structure determination*. Springer, Berlin.

[Mat97] Matsumura, M., and Jinde, Y., "Change of the depth profile of a light emitting zone in organic EL devices with their degradation", *Synth. Met.* 91, 197 (1997).

[McG69] McGlynn, S., Azumi, T., and Kinoshita, M. (1969). *Molecular Spectroscopy of the Triplet State*. Prentice-Hall, New Jersey.

[Mel97] de Mello, J.C., Wittmann, H.F., Friend, R.H., "An improved experimental determination of external photoluminescence quantum efficiency", *Adv. Mat.* 9, 230 (1997).

[Met00] Mettler-Toledo, "Interpreting dsc curves", *User Com* 11, 4 (2000).

[Mon91] Monson, T., Henie, K., William, A., and Mansouri, A., "Tumor-targeted delivery of 8-hydroxyquinoline", *Int. J. Rad. Onc. Biol. Phys.* 6, 1263 (1991).

[Mor97] Mori, T., Obata, K., Miyachi, K., Mizutani, T., and Kawakami, Y., "Fluorescence lifetime of organic thin films alternately deposited with diamine derivative and aluminum quinoline", *Jpn. J. Appl. Phys. 1* 36, 7239 (1997).

[Mur89] Murofushi, E., Kobayashi, S., and Ito, M., "Hydroxyquinoline treatment of aluminum for corrosion resistance and reflectivity", PatentNo JP 01129979 (1989).

[Nan03] Nandagopal, M., Mathai, M., Papadimitrakopoulos, F., Utz, M., "Characterization of isomers in solid aluminum tris-(quinoline-8-olate) by ^{27}Al NMR", Proceedings of the MRS spring meeting, in press (2003).

[Nai93] Naito, K., and Miura, A., "Molecular design for nonpolymeric organic dye glasses with thermal stability: Relations between thermodynamic parameters and morphous properties", *J. Phys. Chem.* 97, 6240 (1993).

[Nak99] Naka, S., Okada, H., Onnagawa, H., Kido, J., and Tsutsui, T., "Time-of-flight measurement of hole mobility in aluminium (III) complexes", *Jpn. J. Appl. Phys.* 38, L1252 (1999).

[Ngu98] Nguyen, T., Jolinat, P., Destruel, P., Clergereaux, R., and Farenc, J. "Degradation in organic light-emitting diodes", *Thin Solid Films* 325, 175 (1998).

[Ohn59] Ohnesorge, W., and Rogers, L., *Spectrochim. Acta, Part A* 15, 27 (1959).

[Pap96] Papadimitrakopoulos, F., Zhang, X.-M., Thomsen, D., and Higginson, K., "A chemical failure mechanism for aluminium(III) 8-hydroxyquinoline lightemitting devices", *Chem. Mat.* 8, 1363 (1996).

[Pap98] Papadimitrakopoulos, F., Zhang, X.-M., and Higginson, K. A., "Chemical and morphological stability of aluminum tris(8-hydroxyquinoline): Effects in light-emitting devices", *IEEE* 4, 49 (1998).

[Per82] Person, W. and Zerbi, G., (1982), *Vibrational intensities in infrared and raman spectroscopy*, Elsevier scientific publishing, New York.

[Pop01] Popovic, Z., Aziz, H., Hu, N.-X., Ioannidis, A., and dos Anjos, P. "Simultaneous electroluminescence and photoluminescence aging studies of tris(8-hydroxyquinoline)aluminum-based organic light-emitting devices", *J. Appl. Phys.* 89, 4673 (2001).

[Pop82] Pope, C.E. and Swenberg, M., (1982), *Electronic processes in organic crystals*. Clarendon Press, Oxford.

[Pro97] Probst, M., and Haight, R., "Unoccupied molecular orbital states of tris(8-hydroxyquinoline)aluminum: Observation and dynamics", *Appl. Phys. Lett.* 71, 202 (1997).

[Rav03] Ravi-Kishore, V., Narasimhan, K., and Periassamy, N., “On the radiative lifetime, quantum yield and fluorescence decay of Alq_3 in thin films”, Phys. Chem. Chem. Phys. 5, 1386 (2003).

[Rho61] Rhodes, W., “Hypochromism and other spectral properties of helical polynucleotides”, *J. Am. Chem. Soc.* 83, 3609 (1961).

[Rie69] Rietveld, H., “A profile refinement method for nuclear and magnetic structures”, *J. Appl. Cryst.* 2, 65 (1969).

[Sap01] Sapochak, L. S., Padmaperuma, A.,Washton, N., Endrino, F., Schmett, G. T., Marshall, J., Fogarty, D., and Forrest, S.R., “Effects of systematic methyl substitution of metal (III) tris(n-methyl-8-quinolinolato) chelates on material properties for optimum electroluminescence device performance”, *J. Am. Chem. Soc.* 123, 6300 (2001).

[Sch01] Schaer, M., Nüesch, F., Berner, D., Leo, W., and Zuppiroli, L., “Water vapor and oxygen degradation light emitting diodes”, *Adv. Funct. Mater.* 11, 116 (2001).

[Sch91] Schmidbaur, H., Lettenbauer, J., Wilkinson, D. L., and Kumberger, O., “Modellsysteme für die Gallium-Extration, Struktur und Moleküldynamik von Aluminium- und Galliumtris(oxinaten)”. *Z. Naturforsch.* 46b, 901 (1991).

[Sch95] Schmidt, A., Anderson, M., and Armstrong, N., “Electronic states of vapor deposited electron and hole transport agents and luminescent materials for light-emitting diodes”, *J. Appl. Phys.* 78, 5619 (1995).

[Shi97] Shi, J., and Tang, C. W., “Doped organic electroluminescent devices with improved stability”, *Appl. Phys. Lett.* 70, 1665 (1997).

[Smi66] Smirnov, V., and Alfimov, M., “Experimental determination of the coefficient characterizing the probability of the transition with $\Delta m = \pm 2$ for triplet states of organic molecules”, *Kinetika i Kataliz* 7, 583 (1966).

[Stö00] Stößel, M., Staudigel, J., Steuber, F., Blässing, J., Simmerer, J., and Winnacker, A., “Space-charge-limited electron currents in 8-hydroxyquinoline aluminium”, *Appl. Phys. Lett.* 76, 115 (2000).

[Ste02] Steiger, J., Schmechel, R., and von Seggern, H., “Energetic trap distributions in organic semiconductors” *Synth. Met.* 129, 1 (2002).

[Sug98] Sugiyama, K., Yoshimura, D., Miyamae, T., Miyazaki, T., Ishii, H., Ouchi, Y., and Seki, K., “Electronic structures of organic molecular materials for organic electroluminescent devices studied by ultraviolet photoemission spectroscopy”, *J. Appl. Phys.* 83, 4928 (1998). 7

[Sve48] Sveshnikov, B., “On the theory of luminescence quenching in organic phosphors”, *Z. Eksp. Teo. Fiz.* 18, 878 (1948).

[Tan87] Tang, C., and VanSlyke, S., “Organic electroluminescent diodes”, *Appl. Phys. Lett.* 51, 913 (1987).

[Tan89] Tang, C., and VanSlyke, S., “Electroluminescence of doped organic thin films”, *J. Appl. Phys.* 65, 3610 (1989).

[Tan03] Ching Tang and coworkers at Eastman Kodak, private communication, May 2003.

[Tsu98] Tsutsui, T., Tokuhisa, H., Era, M., "Charge carrier mobilities in molecular materials for electroluminescent diodes", *SPIE*, 3281, 230 (1998).

[Tsu99] Tsutsui, T., Yang, M.-J., Yahiro, M., Nakamura, K., Watanabe, T., Tsuji, T., Fukuda, Y., Wakimoto, T., , and Miyaguchi, S., "High quantum efficiency in organic light-emitting devices with iridium-complex as a triplet emissive center", *Jpn. J. Appl. Phys. 2* 38, 1502 (1999).

[Tzo01] Tzolov, M., Brütting, W., Petrova-Koch, V., Mückl, A., Berleb, S., Gmeiner, J., and Schwoerer, M., "Subgap absorption in tris(8-hydroxyquinoline) aluminium", *Synth. Met.* 119, 559 (2001).

[Tzo01a] Tzolov, M., Brütting, W., Petrova-Koch, V., Gmeiner, J., Schwoerer, M., "Subgap absorption in poly(p-phenylene vinylene).", *Synth. Met.* 122, 55 (2001).

[Utz03] Utz, M., Chen, C., Morton, M., and Papadimitrakopoulos, F., "Ligand exchange dynamics in aluminum tris-(quinoline-8-olate): A solution state NMR study", *J. Am. Chem. Soc.* 125, 1371 (2003).

[Utz03a] Utz, M., Nandagopal, M., Mathai, M., and Papadimitrakopoulos, F., "Characterization of isomers in aluminum tris(quinoline-8-olate) by one-dimensional ^{27}Al nuclear magnetic resonance under magic-angle spinning", *Appl. Phys. Lett.*, 83, 4023 (2003)

[Vis69] Visser, J.W., "A fully automatic program for finding the unit cell from powder data", *J. Appl. Cryst.* 2, 89 (1969).

[Wat01] Watanabe, T., Nakamura, K., Kawami, S., Fukuda, Y., Tsuji, T.,Wakimoto, T., Miyaguchi, S., Yahiro, M., Yang, M., and Tsutsui, T., "Optimization of emitting efficiency in organic LED cells using Ir complex", *Synth. Met.* 122, 203 (2001).

[Wil72] Williams, D., Metals, ligands and cancer", *Chem. Rev.* 72, 203 (1972).

[Wen03] Wendland, O., (2003). *Triplett-Zustände aromatischer Moleküle in organisch-anorganischen Schichtkristallen - optische Spektroskopie und Nullfeld-ODMR.* PhD thesis, University of Stuttgart.

[Wun90] Wunderlich, B. (1990). *Thermal Analysis*. Academic Press, San Diego.

[Yan02] Yan, L., and Gao, Y., "Interfaces in organic semiconductor devices", *Thin Solid Films* 417, 101 (2002).

[You82] Young, R., and Wiles, D., "Profile shape functions in Rietveld refinements", *J. Appl. Cryst.* 15, 430 (1982).

[You93] Young, R. (1993). *The Rietveld Method.* Oxford University Press, New York.

Acknowledgements

First of all, I am particularly grateful to Prof. Dr. Markus Schwoerer, in whose department I have enjoyed working over the last four years. Not only did he provide essential advice in all questions concerning physics, but he also displayed constant interest in my personal progress.

Special thanks go to Prof. Dr. Tetsuo Tsutsui at the Kyushu University, Japan, who gave me the opportunity to work in his laboratory for two years before I started with my work on Alq_3. He has significantly influenced my way of thinking (and even my way of life), and both his personality and his methods of science have left a lasting impression on me. As a physicist, my focus was on the characterization of devices (at that time based on polymers), and thanks to him I further recognized the importance of the specific properties of the materials used in these devices. Thus, he motivated me to write this book.

This work would not have been possible without the countless long, intensive and helpful discussions with Prof. Dr. Wolfgang Brütting. I would like to express my warmest thanks to him for always being available for questions or advice, and I valued his help and friendship greatly.

It was a great pleasure for me to work in a highly motivated and active team, and I am pleased to acknowledge my colleagues Dr. Marian Tzolov, Dr. Stefan Berleb, Stefan Forero-Lenger, Thomas Stübinger, Anton G. Mückl and last but not least my diploma student Christoph Gärditz.

Another member of this team, but nevertheless someone special, is our lonely chemist among all the physicists: Jürgen Gmeiner. I enjoyed very much our cooperation in preparation of the Alq_3 samples, and whenever possible he demonstrated his profound knowledge in his favorite field of Bavarian culture.

This interdisciplinary work that combined experimental methods from chemistry and physics was only possible in cooperation with the Department of Inorganic

Chemistry I in Bayreuth and the MPI in Stuttgart, especially for structural investigations. In particular I would like to mention Dr. Wolfgang Milius, Prof. Dr. Harald Hillebrecht in Bayreuth and Dr. habil. Robert E. Dinnebier in Stuttgart, and express my gratitude for their intensive and successful cooperation.

Many thanks to Dr. Markus Braun, Dr. Oliver Wendland, Dr. Jost-Ulrich von Schütz and Prof. Dr. Hans Christoph Wolf from the Institute of Physics III at the University of Stuttgart for their enormous help and their great hospitality during my stays in Stuttgart.

I am pleased to acknowledge my former advisor Dr. habil. Walter Rieß for giving me the opportunity to obtain insight into industrial research during my stay at the IBM research laboratory in Rüschlikon, Switzerland.

Most of all, I am grateful to my parents, who have supported me in all that I have done throughout my life.